ACPL ITEM
DISCARDED

```
y                    2250818
542
An2b
Anderson, Curtis B.
Basic experimental chemistry:
a laboratory...
```

**DO NOT REMOVE
CARDS FROM POCKET**

ALLEN COUNTY PUBLIC LIBRARY

FORT WAYNE, INDIANA 46802

You may return this book to any agency, branch,
or bookmobile of the Allen County Public Library.

BASIC EXPERIMENTAL CHEMISTRY:
A LABORATORY MANUAL FOR BEGINNING STUDENTS

C. B. ANDERSON
University of California, Santa Barbara

and

J. L. HAWES
Dillard University, New Orleans

THE BENJAMIN/CUMMINGS PUBLISHING COMPANY
Menlo Park, California
Reading, Massachusetts · London · Amsterdam · Don Mills, Ontario · Sydney

Copyright © 1971 by W. A. Benjamin, Inc. Philippines copyright 1971 by W. A. Benjamin, Inc.

All rights reserved. No part of this publication may be reproduced, stored in a retrieval system, or transmitted, in any form or by any means, electronic, mechanical, photocopying, recording, or otherwise, without the prior written permission of the publisher. Printed in the United States of America. Published simultaneously in Canada. Library of Congress Catalog Card No. 72-85386.

ISBN 0-8053-0222-0
MNOPQRST-AL-898765432

INTRODUCTION

2250818

Chemistry often is described as the study of matter (or substances) and the changes it undergoes. Knowledge of matter and of the energy involved in changes of matter is obtained by observation of nature. Although observations are made outside of experiments, observations under the controlled conditions of an experiment are most valuable. On the basis of the observed data, attempts are made to describe, explain, and understand the observations with general rules, especially with mathematical expressions. Because chemistry is an experimental and quantitative science, it is not possible to understand or appreciate chemistry without an introduction to quantitative experimental work in the laboratory. The objective of this chemistry laboratory course is to familiarize the beginning student with basic laboratory procedures, and to develop the necessary laboratory techniques to use such methods, as well as to demonstrate experimental facts. Students will work individually whenever possible.

It is hoped that this course will teach the student to keep a research notebook and will develop the ability of the individual student to report the essential results of an experiment. For these reasons, the student is required to keep a laboratory notebook (no laboratory report sheets are included in this book). The notebook (or separate report) will be turned in to the instructor for grading a specified time after the laboratory class. The instructor will return the graded notebooks (or reports) at the next laboratory class.

In preparation for the laboratory class, the student should read the experiment procedure with two questions in mind: (1) What is the purpose of the experiment? and (2) What are the quantitative principles involved (or, how are the calculations to be done)?

Equipment and supplies needed are itemized at the beginning of each experiment. The chemicals and expendable items are listed in a separate column with the quantity needed per student (if the experiment is to be performed individually). Where only small amounts are needed, waste can be reduced by supplying the liquids in dropper or plastic squeeze bottles. Otherwise, approximately double quantities are needed because they cannot be measured easily from the reagent bottles and cannot be returned to the reagent bottles. Instructions for preparation of solutions are given in Appendix 2. Equipment requirements per student are given in Appendix 3.

The time required for an average student to complete the laboratory work is included for each experiment. In general, the slowest students are able to finish in 50% more time, that is, within three hours (except, of course, those experiments that are designed for more than one period). The time required depends on the number of students per balance, however.

A set of questions, intended as a study aid, is included at the end of most experiments. These questions review not only important principles in the assignment, but also important considerations in the experimental procedures. The questions are followed by some problems, which are more mathematical and range farther afield.

The historical notes preceding many of the experiments in this book are included for the general education and amusement of the student. It is undoubtedly true that the history of chemistry cannot be understood without first understanding the chemistry, but it is hoped that a little of the historical context of an experiment or principle may make the assignments more interesting (the student is not responsible for this material). The historical material is documented by references included at the end of the book. For further biographical or historical information, the student is urged to consult a standard encyclopedia. If the student should want to track down papers published before 1900, *The Royal Society Catalogue of Scientific Papers*, published by the Royal Society of London, will be found very useful.

The introductory notes present theoretical background, which may not be covered by most standard textbooks, in a way applicable to the assignments.

In some experiments, references are made to some current textbooks containing discussions of the background of the experiment. For other experiments, particularly those dealing with stoichiometry, gas laws, atomic weights, and so on, no references are given. In these cases, if review or reference is needed, the index of the textbook should be consulted. The textbooks referred to are as follows.

F. Brescia, J. Arents, H. Meislich, and A. Turk, *Fundamentals of Chemistry* (Academic Press, New York, 1966).

T. L. Brown, *General Chemistry* (Charles E. Merrill, Columbus, Ohio, 1968), 2nd ed.

R. E. Dickerson, H. B. Gray, and G. P. Haight, Jr., *Chemical Principles* (W. A. Benjamin, New York, 1970).

G. S. Hammond, J. Osteryoung, T. H. Crawford, and H. B. Gray, *Models in Chemical Science* (W. A. Benjamin, New York, 1971).

B. H. Mahan, *College Chemistry*, 1966, or *University Chemistry*, 2nd ed., 1969 (Addison-Wesley, Reading, Massachusetts).

W. L. Masterton and E. J. Slowinski, *Chemical Principles* (W. B. Saunders, Philadelphia, 1969), 2nd ed.

L. Pauling, *College Chemistry* (W. H. Freeman, San Francisco, 1964).

M. J. Sienko and R. A. Plane, *Chemistry: Principles and Properties* (McGraw-Hill, New York, 1966).

CONTENTS

INTRODUCTION		iii
Laboratory practices and operations		1
0.1	The notebook and report	1
0.2	Safety in the laboratory	2
0.3	Basic laboratory equipment	3
0.4	Weighing	3
	0.4-1 The two-pan analytical balance	3
	0.4-2 The single-pan substitution balance	13
0.5	Measuring liquids by volume	15
	0.5-1 Use of the pipette	15
	0.5-2 Use of the burette	16
	0.5-3 Calibration of volumetric glassware	16
	0.5-4 Care and cleaning of volumetric glassware	16
0.6	The Bunsen burner	17
	0.6-1 Heating a test tube	18
	0.6-2 Heating a beaker or flask	18
0.7	Glassworking	19
	0.7-1 Cutting glass tubing	19
	0.7-2 Fire polishing	19
	0.7-3 Bending	20
	0.7-4 Drawing out	20
	0.7-5 Inserting glass tubing or thermometers	20
0.8	Filtration	21
	0.8-1 Analytical filtering	21
	0.8-2 Filtering in synthetic work	23
0.9	Error analysis	24
	0.9-1 Maximum error	24
	0.9-2 Relative error	25
	0.9-3 Average deviation	27
0.10	Slide rule	27

EXPERIMENTS		29
1.	Weighing an unknown with the two-pan analytical balance	31
2.	Gravimetric determination of water	35
3.	Gravimetric determination of total residue of dissolved solids in water	39
4.	Analysis of silver-copper alloy	43

5. The atomic weight of chlorine, and the gravimetric analysis of silver or chlorine as silver chloride — 47
   - Procedure A: Atomic weight by the dry method — 49
   - Procedure B: Atomic weight by the wet method — 50
   - Procedure C: Determination of chloride in a water-soluble chloride or aqueous solution — 52
6. Heat capacity and heat of fusion — 57
   - Part 1: Heat of fusion of ice — 58
   - Part 2: Heat capacity of a metal — 60
7. Molecular weights by vapor density — 63
8. Constant volume gas thermometer — 69
   - Part 1: Measurement of atmospheric pressure — 70
   - Part 2: Calibration of the gas thermometer — 70
   - Part 3: Measurement of the pressure of the gas at the sublimation temperature of carbon dioxide — 72
9. Electrolysis of copper; the faraday — 77
10. Determination of Avogadro's number — 83
11. The combining weight of a metal by displacement of hydrogen — 89
12. The combining weight of a metal by reduction of the oxide — 93
13. Volumetric analysis: titration of acids and bases — 97
    - Part 1: Standardization of the sodium hydroxide titrant — 97
    - Part 2: Determination of the concentration of an acid solution — 98
    - Part 3: Determination of the equivalent weight of an acid — 99
14. Volumetric analysis: potassium permanganate — 103
    - Part 1: Preparation and standardization of $0.2N$ potassium permanganate — 103
    - Part 2: Determination of hydrogen peroxide concentration — 104
    - Part 3: Determination of the percentage of oxalate — 104
15. Complex ions of cobalt — 107
    - Part 1: Preparation of hexaamminecobalt(III) chloride — 111
    - Part 2: Preparation of chloropentaamminecobalt(III) chloride — 112
    - Part 3: Analysis for chloride ion — 113
    - Part 4: *trans*-Dichlorobisethylenediaminecobalt(III) chloride — 114
    - Part 5: *cis*-Dichlorobisethylenediaminecobalt(III) chloride — 115
    - Part 6: Isomerization of *cis*-[Co(en)$_2$Cl$_2$]Cl to *trans*-[Co(en)$_2$Cl$_2$]Cl — 115
16. Potassium trioxalatoaluminate trihydrate — 121
    - Part 1: Preparation of potassium trioxalatoaluminate trihydrate — 122
    - Part 2: Standardization of $0.02M$ potassium permanganate — 122
    - Part 3: Analysis of oxalate — 122
17. Preparation of urea — 125
18. Disodium hydrogen pyrophosphate — 129
    - Part 1: Preparation of sodium pyrophosphate decahydrate — 130
    - Part 2: Preparation of disodium hydrogen pyrophosphate — 130
    - Part 3: Standardization of $0.1N$ NaOH — 131
    - Part 4: Standardization of $0.2N$ KSCN — 131
    - Part 5: Analysis of the disodium hydrogen pyrophosphate — 131

Contents ix

19.	Heat of reaction and heat of solution	135
	Part 1: Heat capacity of the calorimeter	135
	Part 2: Heat of neutralization	135
	Part 3: Heats of solution	136
20.	Molecular weights by freezing point depression	141
21.	Homogeneous equilibrium; the hydrolysis of ethyl acetate	147
	Part 1: Preparation of solutions	147
	Part 2: Analysis of method A or B, one week later	149
22.	Conductance of electrolytic solutions	155
	Part 1: The conductance of acetic acid	160
	Part 2: Change in conductance with acid-base neutralization	161
23.	Indicators and pH	163
24.	The tetraamminecopper(II) cation	169
	Procedure A: Using visual comparison	172
	Procedure B: Using a photometric colorimeter	173
25.	The solubility of silver acetate	179
26.	Inorganic qualitative analysis	185
	Part 1: Group I, Ag^+, Pb^{2+}	187
	Part 2: Group II, Hg^{2+}, (Pb^{2+}), Cu^{2+}, As^{3+}	189
	Part 3: Group III, Fe^{2+}, Ni^{2+}, Al^{3+}	191
	Part 4: Analysis of the unknown for Groups II and III	191
	Part 5: Calcium and ammonium ions	193
	Part 6: Anions, Cl^-, SO_4^{2-}, NO_3^-, CO_3^{2-}, HCO_3^-	194
	Part 7: Qualitative analysis of water	195
27.	Electrochemical cells	199
	Part 1: The Daniell cell	203
	Part 2: Solubility product of cupric hydroxide	205
	Part 3: Formation constant of tetraamminecopper(II) cation	206
28.	Chemical kinetics; decomposition of hydrogen peroxide	211

APPENDIXES		219
1.	References	221
2.	Preparation of solutions	223
3.	Equipment requirements	227
4.	Sample report for Assignment 8	229

LABORATORY PRACTICES AND OPERATIONS

0.1 THE NOTEBOOK AND REPORT

The notebook required is a composition book, about 8 × 10 inches, with a sewn binding and with quadrille ruled paper inside. No other notebooks will be allowed.

Write your name and laboratory section number on the cover of the notebook.

Number every page in the notebook consecutively, beginning at the first page, with ink. *No* pages are to be ripped out of the notebook.

Page 1 is reserved for a table of contents. The assignments will be listed here with their page numbers so that the instructor can find the work.

The assignment number and the date of the work should appear at the top of every page containing either raw data or the report of results.

The raw laboratory data should be written directly in the notebook. It is perfectly good technique to cross out mistakes, and so forth. It is better to record data as neatly and orderly as possible, but this is a secondary consideration. Writing data on stray sheets of paper is bad laboratory technique and will be graded accordingly. One page should be sufficient to record the raw data.

On a page separate from the raw data, preferably the next page, a report of the experimental results should be given. The instructor may prefer that the report be written and submitted separately from the notebook. In either case, the report should contain the following items: the *date* of writing, the assignment *number*, and the *title* or purpose of the experiment. Then give a brief description of the experimental procedure in a few sentences. Do not copy the procedure that is in the book; condense the procedure to a few brief sentences. This should be followed with a clear statement of the experimental results, in grammatical sentences, and a clear statement of the methods of calculation. Data should be presented in the report in neat tables. Also include drawings or answers to questions when they explicitly are required. Always discuss sources of error in your measurements, qualitatively, if not quantitatively.

The report of results and sample calculations should be done first on scratch paper. When they are in coherent form, they should be copied neatly in ink into the notebook (or recopied onto separate sheets of paper for the separate report). A sample notebook page of raw data and a sample report for Experiment 8 are given in Appendix 4. Notice that although there are no headings "purpose," "procedure," and so on, each of the points above has been covered.

0.2 SAFETY IN THE LABORATORY

The laboratory is a dangerous place. To avoid injury to yourself or others, some rules must be made and followed. Furthermore, you must keep in mind some of these potential dangers when you are working in the laboratory.

All students who do not wear prescription spectacles are required to wear safety glasses. Eye protection is required at all times inside the laboratory. You have only one pair of eyes, and accidents are not announced ahead of time. This is considered a part of laboratory technique and is consequently a part of the laboratory grade.

Even though you are wearing glasses, if a chemical should splash into your eye, wash the eye with a large amount of water from an emergency eye-wash fountain (if available), or from a hose attached to the water faucets in the laboratory. This should be done immediately, and the accident should be reported to the instructor.

If chemicals are spilled on your body (or soaked through clothing), wash them off immediately with a large amount of water, using a safety shower or spray, or a hose attached to a water faucet for this purpose.

Report all accidents to the instructor. He will give further instructions.

Know where the fire extinguisher is, and the location of the fireshower and other safety equipment. The instructor will describe how to use the equipment.

Clean up chemical spills immediately.

No unauthorized experiments may be performed.

Follow experimental instructions carefully. Always be sure that you have the right reagent bottle. A serious accident, possibly an explosion, could result from using the wrong reagent.

Most chemicals are poisonous to some degree. Do not taste or smell chemicals unless instructed to do so, and even then do so cautiously. There should be no eating, drinking, or smoking inside the laboratory for safety reasons. It is risky to drink from a laboratory beaker. It is advisable to wash your hands after the laboratory class.

Inserting glass tubing or thermometers into rubber stoppers can be dangerous. Always lubricate the stopper hole with glycerol or water. Always wrap the glass tube with a cloth so that if it breaks, the tube is less likely to gouge your hand. The tube is less likely to break if grasped close to the stopper, and if it is inserted with a twisting motion. See Section 0.7 on glassworking for further advice.

Do not add water to small volumes of concentrated sulfuric acid. Also, do not add concentrated sulfuric acid to small volumes of water. If this is done, the amount of heat generated will be very large, and may cause boiling with consequent spattering. Sulfuric acid is very corrosive, especially when hot. It is safe to add concentrated sulfuric acid to a large volume of water.

0.3 BASIC LABORATORY EQUIPMENT

Some basic laboratory equipment often used in the first-year general chemistry course is shown in Figures 0-1 through 0-5. The illustrations will be helpful in identifying apparatus.

0.4 WEIGHING

Several kinds of balances are used to weigh substances in the chemical laboratory (see Figures 0-6 through 0-8). Which balance is used depends on how accurate the weighing must be and how heavy the material to be weighed is. A platform balance (Figure 0-6), or a triple beam balance (Figure 0-7) can weigh to about 0.1 g or 0.01 g, respectively. If a precision of 0.001 g or 0.0001 g is needed, an analytical balance must be used, and a special technique employed. The load limit on an analytical balance is usually 160 g (or possibly 200 g). The two-pan analytical balance (Figure 0-8) is discussed in the following section. The single-pan automatic balance (Figure 0-9) is discussed afterward. If the single-pan balance is being used, the student should read the principles of the two-pan balance before reading about the single-pan balance.

0.4-1 THE TWO-PAN ANALYTICAL BALANCE

Principles

The two-pan analytical balance is essentially a simple lever having equal arms (see Figures 0-10 and 0-11). In Figure 0-10, E is the main knife edge upon which the lever (called the beam) rests. E_1 and E_2 are knife edges resting on the beam from which the pans are suspended. P is a pointer fixed to the beam.

From the principles of moments, when the beam is level, as judged by the position of the pointer, then

$$F_1 l_1 = F_2 l_2$$

where F_1 and F_2 are the forces exerted on the arms of the lever, and l_1 and l_2 are the lengths of the arms. The force F is the product of the mass and the gravitational attraction, $F = mg$. So when the beam is level, the mass on pan one is equal to the mass on pan two, if the arms of the lever are equal in length

$$m_1 g l_1 = m_2 g l_2$$
$$l_1 = l_2$$
$$m_1 = m_2$$

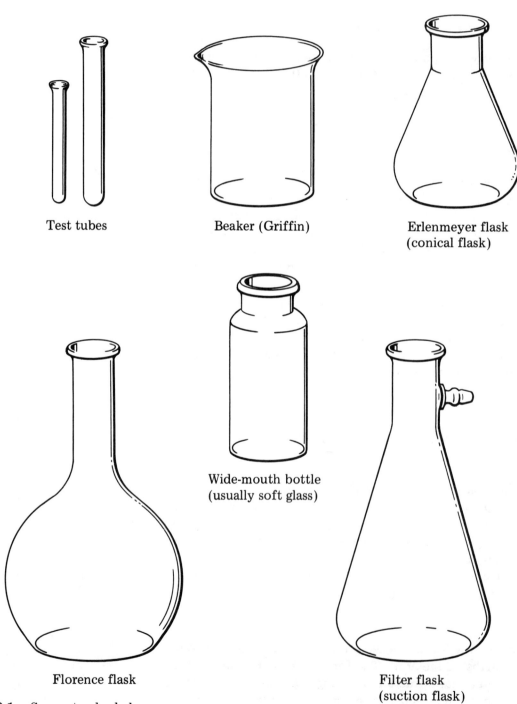

Figure 0-1. Some standard glassware.

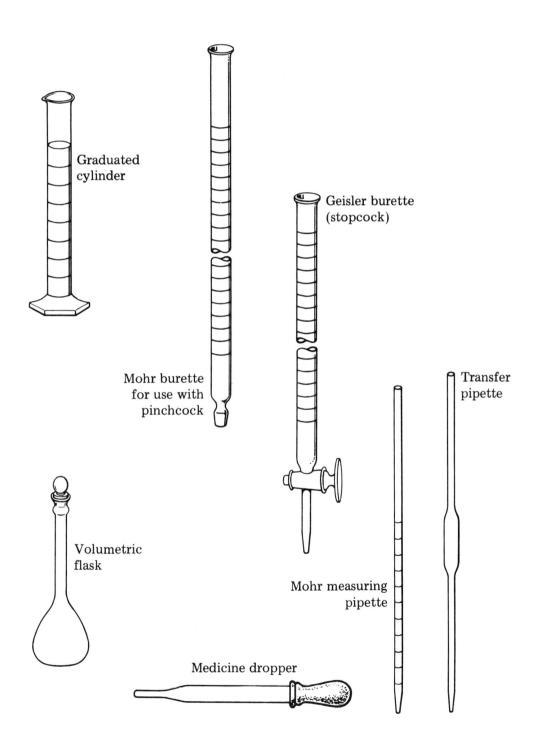

Figure 0-2. Volumetric equipment.

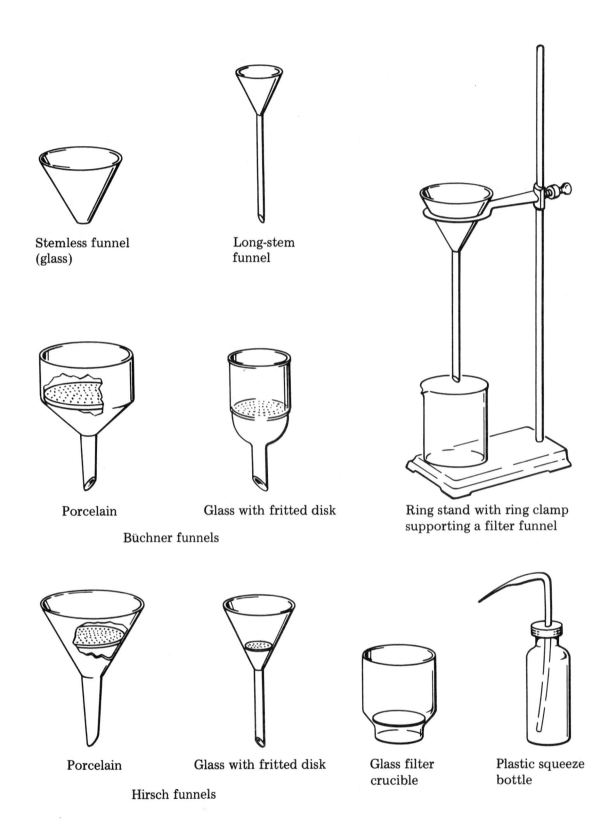

Figure 0-3. Filtering equipment.

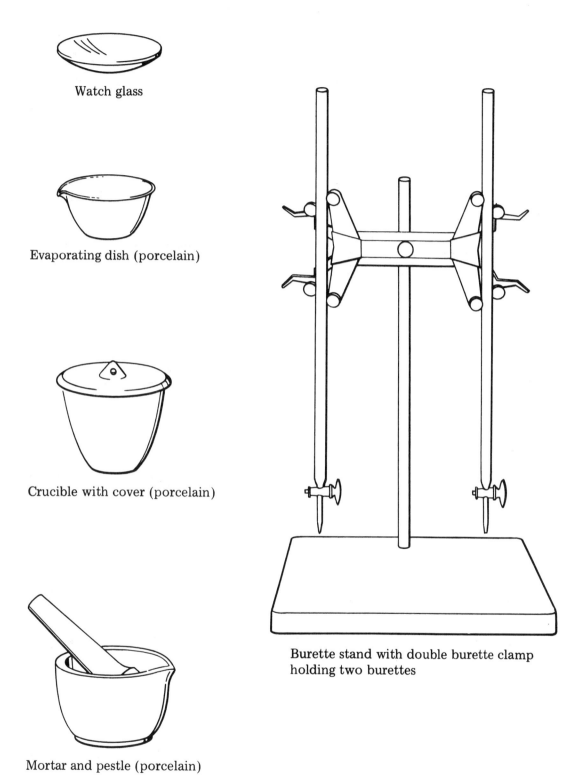

Figure 0-4. Miscellaneous equipment (not to scale).

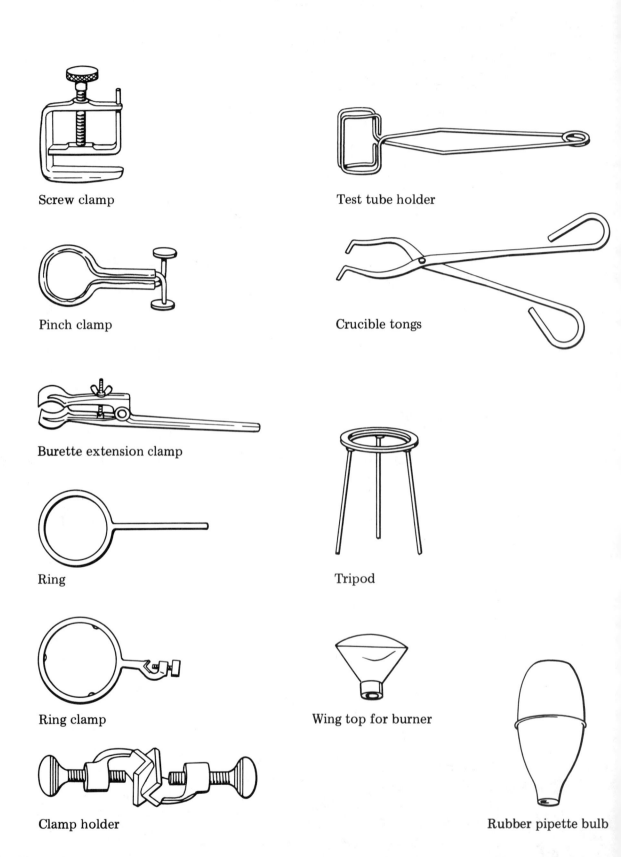

Figure 0-5. Miscellaneous hardware (not to scale).

Weighing 9

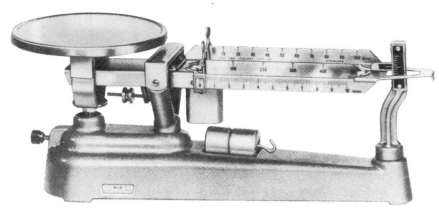

Figure 0-6. Platform triple beam balance.

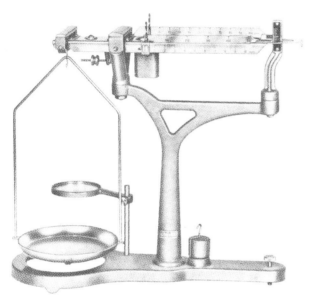

Figure 0-7. Triple beam balance.

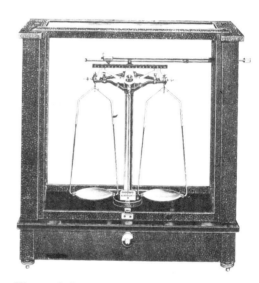

Figure 0-8. Two-pan analytical balance with rider.

Figure 0-9. Single-pan automatic balance.

10 Laboratory practices and operations

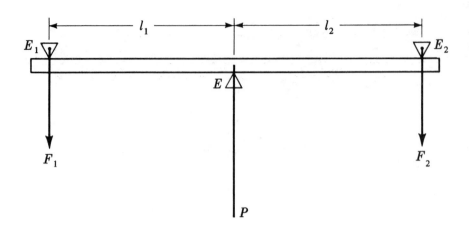

Figure 0-10. Lever with equal arms (compare this diagram with the drawing of the balance beam in Figure 0-11).

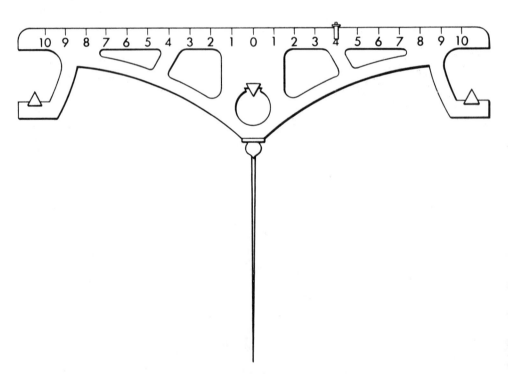

Figure 0-11. The beam of a rider balance with rider at 4 mg.

Notice that the balance compares masses (m). The weight of a certain object will vary, depending on the gravitational attraction; consequently, the weight (F) of this object will be smaller in the mile-high city of Denver than in Santa Barbara on the beach. In everyday speech, this distinction between mass and weight is ignored because in any locality *g* is a constant, thus weight is proportional to mass. If this is understood, it is permissible to use the word *weight* as another word for *mass*, although strictly speaking, they are not the same.

Care of the balance

Whenever objects or weights are placed on or taken from the balance pans, the steel knife edges must be raised off the agate plates to prevent damage and wear. This is accomplished by turning a knob (on the front of the balance case) that raises the "beam arrest," which in turn raises the beam and the stirrups, from which the pans are suspended.

The pan arrests support the pans and keep them from swinging. They are controlled by a button on the front of the balance case. When the beam arrest is up, the pans should be arrested also.

Because the balance compares the mass of the object to be weighed with the mass of some brass weights, it is obvious that the weights must be kept clean. This means that weights must not be touched with fingers. Fingers are moist and greasy even on clean hands! Weights must be handled *only* with forceps.

There is a wire rider on the top edge of the beam that is used for weights smaller than 10 mg (Figure 0-11). Use the rider rod to move the rider. The rider must not be touched with fingers. If it is knocked off, it must be handled with forceps only, and care must be taken not to bend it.

Only dry solid objects may be weighed directly on the balance pans. Powders and liquids must be weighed in containers. The pans must be kept clean. If a bit of dust or powder is on the pan, it should be brushed off with a camel's hair brush.

Operation of the balance

Finding the zero point

Sit down in front of the balance. The balance door should be closed. Look at the pans to see that they are clean. Check to see if the beam arrest and pan arrests are up. Move the rider to the 0-mg position.

Now lower the beam arrest and lower the pan arrests. The pointer fixed to the beam will oscillate back and forth slowly. If it doesn't, the balance door may be raised and air *gently* wafted against the pan with the hand. Care must be taken not to hit the instrument, or to stir up the air too vigorously. Remember to close the balance door again. The pointer should oscillate back and forth with an initial amplitude equal to about five divisions to the right or left. The amplitude of the oscillations will decrease with time due to friction in the bearings and air damping. Let the pointer make three swings. This will allow air currents inside the balance case to subside. Then read the maximum displacement of the pointer from the scale behind the pointer (Figure 0-12). The maximum displacement of an odd number of swings is taken. For example, the pointer swings to +1.9 (1.9 divisions to the right of the center line on the scale) and swings back to −2.8 (2.8

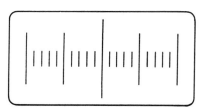

Figure 0-12. Pointer scale.

divisions to the left). Then it swings to +1.7, to −2.7, and then to +1.6. After making these readings, release the pan arrests when the pointer is near the middle of the scale, and then raise the beam arrest.

Take the average of the swings to the right, $1/3(+1.9 + 1.7 + 1.6) = +1.7$, and the average of the swings to the left, $1/2(-2.8 - 2.7) = -2.75$, and then take the median of the two averages. This value (−0.5 division to the left) is the rest point. Since there is nothing on the pans, this rest point (−0.5) is the zero point. The zero point may change from day to day, or even more often, due to changes in temperature, and so on. The zero point is not necessarily at the center of the pointer scale, but it should be within one or two scale divisions of the center. If it isn't, ask the instructor to adjust the balance. The rest point of the balance, when the beam is not swinging, may be considerably in error because of friction in the bearings.

Sensitivity of the balance

Set the rider at 3 mg with the rider control. Do this slowly and carefully. Now record five swings of the balance pointer after letting three swings go by: −0.5, −6.5, −0.6, −6.4, −0.7. The rest point is −3.53 divisions to the left. Therefore, 3 mg on the rider shifts the rest point from −0.5 to −3.53, a shift of −3.03 divisions on the scale. The sensitivity of the balance is then $3.03/3 = 1.0$ division per milligram, or 1.0 mg per division. The sensitivity is ordinarily slightly different at different loadings of the balance. That is, the sensitivity may be slightly different when a 1-g object is being weighed and when a 20-g object is being weighed. This difference should be checked. Obviously, different balances will have different sensitivities.

The weighing

With the beam arrest up, place the object to be weighed (for example, the 25-ml flask) on the left-hand pan of the balance. The object is counterbalanced by putting weights on the right-hand pan, beginning with the largest weights. Therefore, using forceps, place the 50-g weight on the right pan. Now lower the beam arrest *partly* to see which way the pointer moves. If the pointer moves to the left, the weight is too large. For a 25-ml volumetric flask, this will be the case. Raise the beam arrest again and remove the 50-g weight with the forceps. Next try the 20-g weight. If the object is heavier than 20 g, add the other 20-g weight, and test that by lowering the beam arrest partly. Eventually, the weights will be adjusted so that the pointer does not move when the beam arrest is lowered. Then depress the pan arrests and add fractional weights to counterbalance the object as closely as possible. Whenever weights of 1 g or larger are added, the beam arrest must be up. Whenever any weights are taken off the pan, the beam arrest must be up.

This weighing will now be to 10 mg. Further counterbalancing can be achieved by using the rider. When the object has been counterbalanced by weights and the rider so that the balance pointer will swing freely and the rest point is on the scale, read five swings after letting three go by: +3, −3, +2.8, −2.8, +2.6. The rest point is −0.05 division. The weights on the pan were, for example, 20 g + 500 mg + 200 mg and the rider is at +4 mg. This is a total of 20.704 g.

However, the rest point (−0.05) is not exactly the zero point of the balance (−0.50). To evaluate this difference, we need the sensitivity at this load (roughly 20 g). Therefore, the rider is shifted 2 mg to the right and the rest point taken: +1.0, −4.6, +0.8, −4.4, +0.6. The rest point is now −1.85 division, which is a change of −1.85 − (−0.05) = −1.8 divisions for a change of 2 mg, or 1.1 mg per division. Consequently, the difference between the rest point (−0.05) and the zero point (−0.5), which is 0.45, is equivalent to 0.45 division (1.1 mg division^{-1}) = 0.5 mg. Because the rest point (−0.05) is to the right of the zero point (−0.5), the object was 0.5 mg heavier than the 20.704 g of weights on the other pan. Therefore, the weight of the object is 20.704 g. + 0.0005 g = 20.7045 g.

0.4-2 THE SINGLE-PAN SUBSTITUTION BALANCE

The double-pan balances just described weigh by addition of weights to the weight pan, which counterbalances the object on the other pan. In contrast, the modern single-pan balance weighs by substitution of weights. The single-pan substitution balance employs a beam, which is a lever with unequal arms (see the simplified diagram in Figure 0-13). The beam has two knife edges. The center knife edge acts as the fulcrum for the lever. The pan for the object to be weighed and a weight carrier are suspended from the other knife edge. The pan and weight carrier are counterbalanced by a counterweight fixed to the other end of the beam.

The weight carrier contains a set of weights totaling 160 g (or possibly 200 g). The weights may be rings, hooks, bars, or so on, in various types of balances. When an object on the pan is to be weighed, the operator dials a knob connected mechanically to fingers that lift an

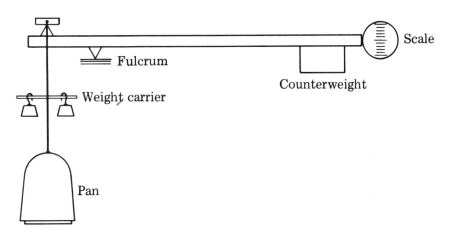

Figure 0-13. Single-pan balance (simplified).

appropriate number of weights from the weight carrier until the weight of the object and the weights in the weight holder are once again the same as the total weights (160 g). That is, if a 50-g object is to be weighed, weight knobs are dialed so that 50 g of the weight set is removed from the weight holder by the mechanical system. Then the total weight of the object and weights still in the weight holder is 160 g. When this is exactly so, the beam will return to the same place of balance when it is released. Notice that when the beam is balanced, it always has the same load (160 g); thus, the sensitivity is essentially constant.

Weight differences smaller than the smallest weight in the weight carrier are measured by the deflection of the beam from the original zero point. This is measured by an optical system that magnifies the image of a small scale attached to the beam onto a ground glass. The optical system differs somewhat among various manufacturers' balances. A damping vane also is attached to the beam to bring the beam to rest quickly.

There are several advantages of the single-pan substitution balance over the two-pan balance:

1. Errors due to differences in the length of the beam arms in a two-pan balance are not possible in the single-pan system.
2. Two knife edges instead of three reduce friction.
3. Substitution weighing results in constant sensitivity.
4. In substitution weighing, the weights are not handled, thus preserving their calibrations.
5. Weighing is easier and quicker.

Operation of the balance

The exact operating instructions vary somewhat among various makes of balances. The instructor will give detailed instructions. In general, the following steps are observed:

1. Check the spirit level on the balance. Adjust the leveling screws until the bubble is within the circle. Changing the leveling screws changes the zero point of the balance.

2. Check the balance zero. With nothing on the pan (clean it with a brush if necessary), the weight knobs at zero, and the balance doors closed, slowly release the beam completely. When the optical scale comes to rest, set the zero of the scale with the zero adjustment knob. Now arrest the beam slowly. The beam arrest knob turns easily. Do not apply force.

3. Put the object to be weighed on the balance pan. Keep hands outside the balance as much as possible to prevent causing heat gradients inside the balance. Gently turn the beam arrest knob to the partial arrest position. Dial weights with the beam in partial arrest until the weight is adjusted within the range of the optical scale. Now the beam may be released fully. Read the weight from the dialed weights and the optical scale. Arrest the beam slowly and gently. Return the weight

knobs to zero. Remove the object from the pan. If anything has been spilled, clean it up immediately or report it to the instructor. Close the balance doors.

Care of the balance

Do not add or remove objects from the pan unless the beam is arrested fully.

Dial weights only when the beam is in partial arrest or full arrest.

Only dry solid objects may be weighed directly on the balance pans. Powders and liquids must be weighed in containers. The pans must be kept clean. If dust or powder is on the pan, it should be brushed off with a camel's hair brush.

0.5 MEASURING LIQUIDS BY VOLUME

Equipment used for measuring the volume of liquids includes volumetric flasks, graduated cylinders, pipettes, and burettes. Volumetric flasks and cylinders are graduated to contain (TC) a certain volume of liquid at a certain temperature. Pipettes and burettes (and sometimes graduated cylinders) are calibrated to deliver (TD) a certain volume of water at some specified temperature. Marks designating the volume usually are etched into the glass. Since water wets glass, the surface of the water in the glass tube of one of these containers looks as is shown in Figure 0-14, where the surface of the water is curved, and is higher at the walls of the tube. The meniscus (the curved surface of the water) should appear just to touch the etched line when viewed from the side with the eye level at the etched mark. Volumetric flasks and transfer pipettes are narrow at the place where the mark is so that errors in getting the meniscus to the line are minimized. Graduated cylinders are not used for precise measure, so they need not be read as carefully. Burettes are used to measure volumes of liquid precisely. A burette is simply a long graduated tube of uniform bore with a stopcock or pinchcock at the end. In order to read the level of a liquid with respect to a line on the burette, hold a white card, which has the lower half blackened, behind the meniscus. The card is shown in Figure 0-15. Hold the card so that the black area just appears to touch the bottom of the meniscus of the liquid. The eye must be level with the meniscus of the liquid to eliminate parallax errors. Then read the top of the black part of the card with respect to the graduations on the burette.

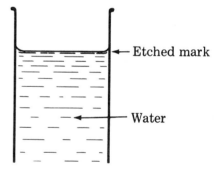

Figure 0-14. Water meniscus.

0.5-1 USE OF THE PIPETTE

The method of using graduates, volumetric flasks, and burettes is fairly obvious. However, pipettes often are used incorrectly and dangerously. Pipettes often are used for drawing up poisonous and corrosive solutions. Therefore, you must *never* use your mouth to supply the suction. Also, your breath would make the pipette dirty. Always use a pipette bulb to draw a liquid into the pipette. Holding the pipette in one hand,

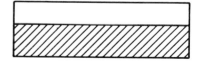

Figure 0-15. Card to aid in reading a burette.

16   Laboratory practices and operations

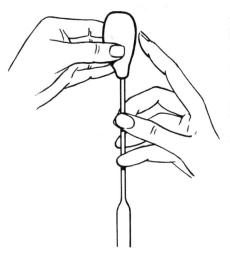

Figure 0-16. Using a pipette bulb.

dip the pipette into the solution to be pipetted, and with the other hand squeeze the pipette bulb. Hold the open end of the pipette bulb tightly to the top of the pipette (Figure 0-16). Loosen your grip on the bulb so that it sucks liquid into the pipette. Do not push the pipette bulb over the end of the pipette (unless the pipette bulb has valves in it). When the liquid is above the mark on the pipette, take away the bulb, and put the index finger of the hand holding the pipette as shown in Figure 0-17. Now carefully let the liquid run out until the meniscus is exactly at the mark. Take the pipette tip out of the liquid, wipe it with a tissue or towel (to take away any drops on the outside), and let it drain into the proper container. Do not force all of the liquid from the pipette. Pipettes are calibrated with the little bit that does not drain out of the pipette taken into account. Simply touch the pipette to the side of the container into which the liquid is being pipetted for about 20 seconds after the liquid has run out.

0.5-2 USE OF THE BURETTE

The burette must be read before and after delivering the liquid desired. The difference between these readings is the volume of liquid delivered. The most convenient container for titration is the Erlenmeyer (conical) flask. When titrating, a right-handed worker probably will find it most convenient to operate the stopcock or pinchcock with his left hand and to swirl the flask with his right hand. The pinchcock should be grasped with the round end toward the palm of the hand. Similarly, the stopcock plug end should be toward the palm of the hand. The liquid should be added drop-by-drop. If the liquid is allowed to run out too fast, the walls of the burette will not drain properly (and also, you may go past the end point).

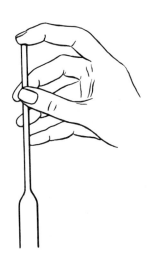

Figure 0-17. Operation of a pipette.

0.5-3 CALIBRATION OF VOLUMETRIC GLASSWARE

The calibration of volumetric equipment usually is carried out by measuring distilled water and weighing it on an analytical balance. From the weight of water and the density of the water at a specific temperature, the volume of the apparatus can be calculated. The density of water at various temperatures is listed in handbooks.

0.5-4 CARE AND CLEANING OF VOLUMETRIC GLASSWARE

Always rinse volumetric glassware three times with distilled water after use. If something dries on the glass, it may be difficult to remove later.

Do not dry calibrated volumetric glassware in an oven. The glass expands and contracts irreversibly and may change the calibration.

Water should wet the glass evenly. If it does not, grease or other dirt must be on the glass. To remove grease, fill the piece of apparatus with hot detergent solution one or more times. Rinse it with distilled water, and let it drain. If this does not remove the dirt, stronger cleaning agents must be used. However, do not use "cleaning solution" (potassium dichromate in concentrated sulfuric acid) without instruction and assistance from your laboratory instructor. It is an exceedingly corrosive liquid, which is why it cleans. To draw detergent solution into a burette, use vacuum or an aspirator, as shown in Figure 0-18.

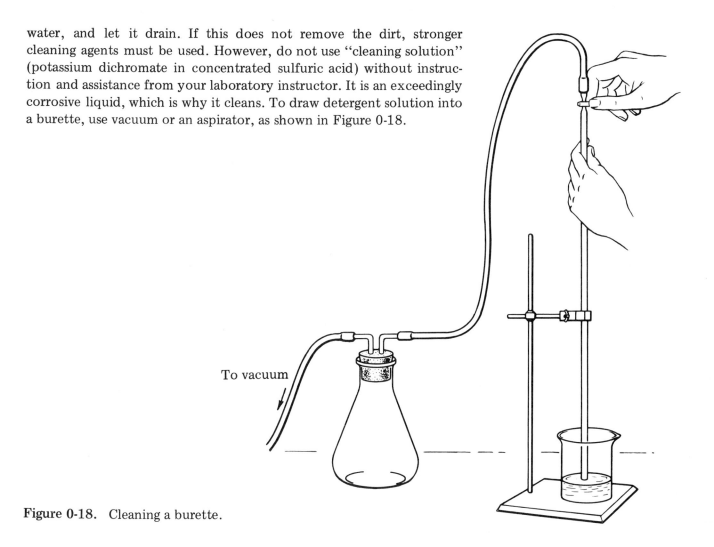

Figure 0-18. Cleaning a burette.

0.6 THE BUNSEN BURNER

Most heating in the first-year laboratory course is done with a gas burner, although an oven may be used in some cases. More advanced laboratory procedures may require electric heating mantles, hot plates, electrically heated baths, oxygen torches, and so on. The gas burner used in this laboratory course probably will be of either the Bunsen or the Tirrill type. The Bunsen burner usually has no gas needle valve and has an air vent that is adjusted by rotating a sleeve. The Tirrill burner, shown in Figure 0-19, has a gas needle valve, and the air is controlled by unscrewing or screwing the top of the burner.

The fuel used in the burner is usually natural gas, which is mostly methane, CH_4. If the air inlet is closed and the gas is lit, the flame will be large and luminous. The light is the radiation given off by hot carbon particles that are burned only partially. This flame is not very hot. If the air control is adjusted so that air is mixed with the gas before it gets to the flame, the flame will become less luminous, and finally blue. When the air is adjusted correctly to give the hottest flame, it will look

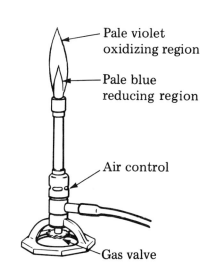

Figure 0-19. Tirrill burner.

more or less as shown in Figure 0-19. The inner cone of the flame is pale blue, and the outer cone is a pale violet. The inner cone contains unburned gas that is hot enough to radiate light. The inner cone is not as hot as the outside cone. The hottest point in the flame is just above the inner cone.

0.6-1 HEATING A TEST TUBE

If a test tube is to be heated with a burner, hold the test tube with a test tube holder at an angle, and heat just below the surface of the liquid, not at the bottom (Figure 0-20). Also, jiggle the test tube a little. If the test tube is heated on the bottom, a bubble may form and violently eject the entire contents of the test tube. This is called bumping, and can cause a serious accident if the test tube is pointed at you or someone else. Therefore, when you are heating a test tube over a burner, do not point it at anyone. If the test tube must be heated just to boiling, it may be more convenient to heat it in a beaker of boiling water.

0.6-2 HEATING A BEAKER OR FLASK

If a beaker or flask of water must be heated, set the beaker (or flask) on a wire gauze that is supported by a ring clamp on a ring stand as in Figure 0-21. For even boiling, it is desirable to add a boiling stone. A boiling stone is a chip of Carborundum, marble, or other porous material that has tiny fissures. The air trapped in these fissures provides a gas phase into which the hot liquid will vaporize readily. Without such a gas phase, the liquid becomes superheated at the bottom of the beaker until it gets so hot that a tiny bubble forms. Then a violently rapid vaporization takes place, suddenly producing a large bubble. This

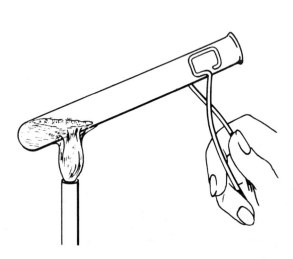

Figure 0-20. Heating and boiling a solution in a test tube.

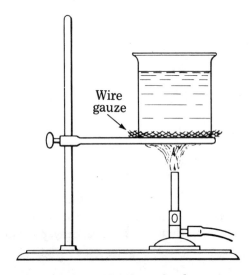

Figure 0-21. Heating a beaker on a ring stand with a Bunsen burner.

is called "bumping." Smooth, unscratched beakers and flasks "bump" very readily.

0.7 GLASSWORKING

0.7-1 CUTTING GLASS TUBING

Lay the tubing flat on the desk. With one stroke of the file, make a single deep scratch at the desired length. Wet the scratch. Grasp the tubing with both hands with your thumbs on the side opposite the scratch. Now push gently and carefully, but firmly, with the thumbs. Also exert slight pressure tending to pull the two pieces of glass apart. If the glass doesn't break, make a deeper scratch and try again. See Figure 0-22.

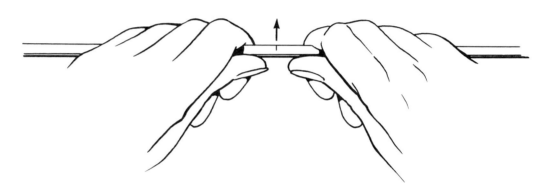

Figure 0-22. Breaking the scratched glass tube.

0.7-2 FIRE POLISHING

The edges of freshly broken glass tubing are sharp. These sharp edges may be removed by rotating the end of the tubing in the Bunsen burner flame until the edges melt slightly. Do not heat so long that the tubing melts together.

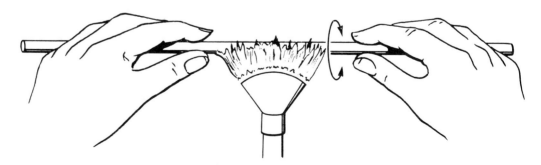

Figure 0-23. Rotation of the glass tube.

0.7-3 BENDING

Put a wing tip on the Bunsen burner. (The burner must be turned off to do this.) Then rotate a piece of glass tubing (slowly and evenly with both hands) in the flame of the wing tip (Figure 0-23). When the glass is soft, remove it from the flame and bend to the angle desired. The glass should bend easily and should not require force. A correct bend is shown in Figure 0-24.

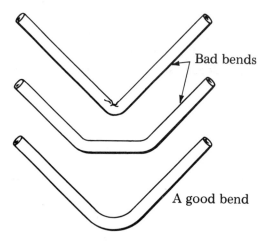

Figure 0-24. Bending of glass tubes.

0.7-4 DRAWING OUT

As for bending, rotate a piece of tubing in the flame of the wing tip until it softens. Then take it out of the flame and pull the two pieces apart. After the glass has cooled, it can be cut to convenient size for eyedroppers. Fire-polish the sharp edges!

0.7-5 INSERTING GLASS TUBING OR THERMOMETERS

Before inserting glass tubing into a cork or rubber stopper, fire-polish the sharp edges of the tubing as described. Lubricate the glass and the hole with glycerol (water is not as good). Cover the glass with a cloth so that if the glass breaks, it will not gouge your hand. Grasp the glass with the towel around it. Insert the glass tube with a twisting motion, holding it close to the stopper. This operation is shown in Figure 0-25.

When taking glass tubing out of a stopper, work glycerol into the holes around the glass before you try to remove the stopper by the usual twisting motion. Cork borers may be used to remove glass tubing from rubber stoppers. Select a cork borer that just barely fits around the tube, and with a twisting motion work it between the rubber stopper and the glass. When the cork borer is all the way through, the glass tube can be taken out easily. See Figure 0-26.

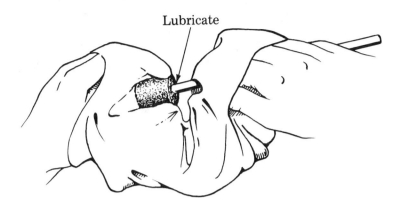

Figure 0-25. Inserting a glass tube into a rubber stopper.

0.8 FILTRATION

Filtration is the method of separating a solid from a liquid by passing the liquid through a porous material. Decantation, and centrifugation followed by decantation, are two other methods of separating solids from liquids. Decantation is a simpler method, consisting merely of pouring off the liquid from the solid. However, since many solids are finely divided, decantation can be difficult. Centrifugation can be used to pack the solid against the bottom of the container, after which decantation is possible. In filtration, the porous filtering material can be cloth (as in industrial filter presses), paper, sintered glass or porcelain, a mat of fiber glass or asbestos, and so on. Filters also can be obtained in a variety of porosities. If the filter has larger pores, the liquid will pass through more easily, and filtration will be faster. However, the larger the size of the pores in the filter, the larger the particles of solid that can pass through. The choice of method of filtration (see the following sections) and the filtering material depend on the application in question.

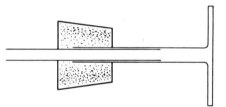

Figure 0-26. Removing a glass tube from a rubber stopper, using a cork borer.

0.8-1 ANALYTICAL FILTERING

In quantitative analysis, filtering usually is done with a long-stem conical funnel into which a cone of filter paper is inserted. In addition to the various porosities of paper available, there are quantitative and qualitative filter papers. Quantitative filter paper is used when the precipitate (solid) that is filtered off must be weighed. In this case, the solid precipitate is dried in the filter paper, the filter paper is burned, and the precipitate with the residue of the paper is weighed. Quantitative paper is manufactured specially so that it leaves a very small and reproducible ash. Since quantitative filter paper costs more than regular filter paper, the qualitative grades are used for all other applications.

The filter paper is folded to fit the conical funnel, as in Figure 0-27. Fold the filter paper in half again. Open the folded filter paper into a cone with the torn-off corner outside. Fit the cone into the

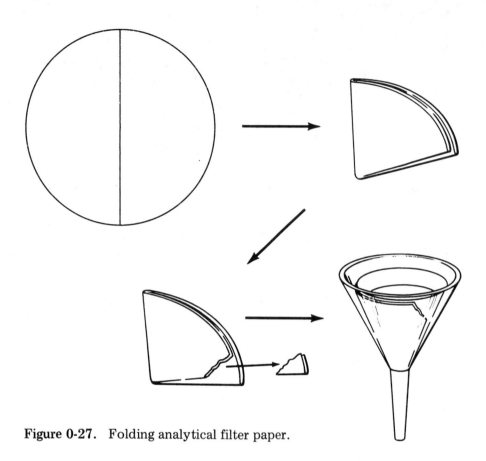

Figure 0-27. Folding analytical filter paper.

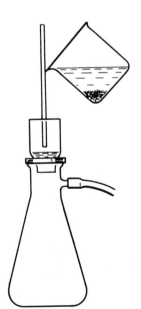

Figure 0-28. Filtering a liquid through a sintered glass crucible with suction.

funnel. Wet the paper with water, and adjust it so that the top edge is wetted tightly to the glass. Add more water so that the stem of the funnel fills with water. If the filter paper is fitted correctly, the filter paper will support a column of water in the funnel stem. The weight of this column of water produces a mild suction that expedites filtration.

For quantitative work, especially where paper is unsuitable because of corrosive liquids or because of its reducing properties on burning, sintered glass funnels and crucibles, or porcelain crucibles with porous bottoms, often are used. These often are used even when paper could be used, because they are convenient, and have a further advantage in that suction can be used to expedite filtration. However, they are expensive.

In quantitative work, the liquid (with the suspended solids) must be poured carefully into the filter funnel so that no solid precipitate is lost. This is accomplished by pouring the liquid down a stirring rod into the funnel. The stirring rod guides the liquid into the funnel. This procedure is illustrated in Figure 0-28. When the liquid has been filtered, the solid is washed into the funnel with a spray of water from a plastic squeeze bottle (Figure 0-29). The stream of water and solid is guided into the filter funnel with a stirring rod held as shown. If some particles of solid will not wash into the funnel, they must be loosened from the glass. A rubber policeman on a stirring rod is useful for this purpose (Figure 0-30).

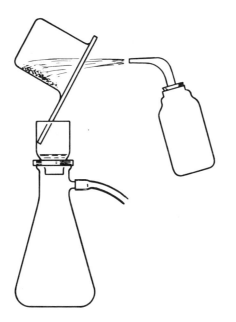

Figure 0-29. Washing the precipitate into the crucible with a squeeze bottle.

Figure 0-30. Stirring rod with rubber policeman.

0.8-2 FILTERING IN SYNTHETIC WORK

In synthetic work, suction filtration using paper filters on Büchner or Hirsch funnels is very common. Büchner and Hirsch funnels usually are made of porcelain. The inside bottom of the funnel is a flat plate with holes in it that supports the filter paper. Büchner and Hirsch funnels also can be obtained in glass with sintered glass disks, although they are expensive. These are used in the same way, but without the filter paper. The porcelain Hirsch funnel is shown in use in Figure 0-31. A rubber ring forms the seal between the funnel and the filter flask, which is connected to the vacuum line or to the aspirator. (A rubber stopper or cork also can be used to fit the funnel to the filter flask instead of the rubber ring.) When using this type of funnel, place the appropriate size filter paper circle in place and turn on the vacuum. Then wet the filter paper with the liquid to be filtered so that the paper is held tightly to the funnel, and solid cannot possibly get under the edge of the filter paper. Then begin filtering.

Solutions of very volatile liquids, such as ether, or hot solutions, are not filtered very conveniently with suction, because the suction may cause excessive evaporation of the solvent, which may cool the solution enough to cause precipitation of the solute. A convenient filter can be made by stuffing a wad of cotton or glass wool in the apex of the cone of a short stem funnel. Another method is the use of the fluted filter paper in a short stem conical funnel. The fluted filter paper has the advantage of a large surface area for filtering that makes it filter faster. The fluted filter often is used for filtering volatile solvents such

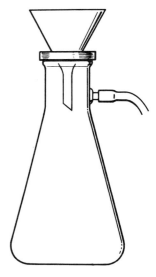

Figure 0-31. Suction filtration with a Hirsch funnel.

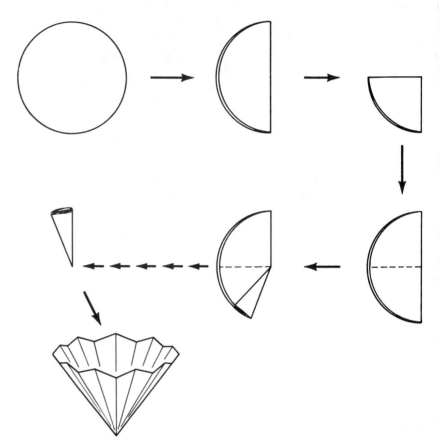

Figure 0-32. Folding a fluted filter.

as ether. To fold the filter paper (Figure 0-32), fold it in half, and then fold it in half again to find the middle. Unfold to half-folded, and fold the half-circle in accordianlike folds. Open the filter, and place it in the funnel.

0.9 ERROR ANALYSIS

A discussion of errors should be included in the report of each experiment that involves a quantitative measurement. This discussion may be limited to an enumeration of the possible causes of error, and their effects on the results. However, for many of the experiments, the limits of error can be estimated quantitatively. Whether you are required to discuss errors quantitatively in your report for a particular experiment will be decided by your instructor.

0.9-1 MAXIMUM ERROR

Generally, the error in a measurement can be assumed to be half the smallest scale division. Thus, for a 1-mg, single-pan automatic balance, the weighings are to ±0.0005 g. For a 50-ml burette with 0.1-ml markings, the error could be ±0.05 ml, although in this case it is possible to read to ±0.02 ml. In any case, the limits of the reading can be estimated easily, so every measured value has an error for which we can estimate error limits. When these measured values are added, subtracted, multiplied, or divided, it is not immediately clear how to combine the errors to arrive at the resultant value and its error. How errors combine in addition, subtraction, multiplication, and division is discussed in the next section. The maximum error represents the most extreme value that the number can have; that is, the true value is certain to be within this limit. "Probable error" is a much smaller quantity, and is arrived at in a more complicated way.

Maximum error in addition or subtraction

The maximum error in addition or subtraction of two numbers A and B, having errors e_a and e_b, can be stated as the sum of the two errors e_a and e_b:

$$(A \pm e_a) + (B \pm e_b) = (A \pm B) \pm (e_a + e_b)$$

$$(A \pm e_a) - (B \pm e_b) = (A - B) \pm (e_a + e_b)$$

This can be seen numerically in

$$(1.00 \pm 0.02) + (1.00 \pm 0.03) = 2.00 \pm 0.05$$

because the extremes are

$$1.02 + 1.03 = 2.05 \text{ and } 0.98 + 0.97 = 1.95$$

Maximum error in multiplication

The maximum error in multiplication is given by the equation

$$(A \pm e_a)(B \pm e_b) \cong AB \pm (Ae_b + Be_a)$$

since

$$(A + e_a)(B + e_b) = AB + Ae_b + Be_a + e_a e_b$$

and $e_a e_b$, the product of two small numbers, is negligible.

0.9-2 RELATIVE ERROR

Relative error is the fraction of error, e_a/A in the value A,

$$\frac{A \pm e_a}{A} = 1 \pm \frac{e_a}{A}$$

In multiplication, the relative error in the product is

$$\left(1 \pm \frac{e_a}{A}\right)\left(1 \pm \frac{e_b}{B}\right) = 1 \pm \frac{e_a}{A} \pm \frac{e_b}{B} + \frac{e_a e_b}{AB}$$

The error-squared term is the product of two small numbers, which will be negligible. Consequently,

$$\left(1 \pm \frac{e_a}{A}\right)\left(1 \pm \frac{e_b}{B}\right) \cong 1 \pm \frac{e_a}{A} \pm \frac{e_b}{B}$$

The maximum limits of error then would be

$$1 \pm \left(\frac{e_a}{A} + \frac{e_b}{B}\right)$$

This relationship is the same as that given for maximum error in multiplication, when it is multiplied by AB:

$$AB\left[1 \pm \left(\frac{e_a}{A} + \frac{e_b}{B}\right)\right] = AB \pm Be_a \pm Ae_b$$

For the maximum error,

$$AB \pm (Be_a + Ae_b)$$

For division it is easier to show the propagation of relative error than the error itself. Division of the fraction $1/1 \pm d$ gives the relationship

$$\frac{1}{1 \pm d} = 1 \mp d + d^2 \mp d^3 + \ldots$$

which approximates

$$\frac{1}{1 \pm d} \cong 1 \mp d$$

if d is small compared to 1 and d^2 consequently is negligible, as will be all higher terms. The propagation of relative error in the division of $A + e_a$ by $B \pm e_b$ is

$$\frac{1 \pm \dfrac{e_a}{A}}{1 \pm \dfrac{e_b}{B}} = \left(1 \pm \frac{e_a}{A}\right)\left(1 \mp \frac{e_b}{B}\right)$$

$$= 1 \pm \frac{e_a}{A} \mp \frac{e_b}{B} - \frac{e_a e_b}{AB}$$

or if the square term can be neglected, the error lies between

$$1 + \frac{e_a}{A} + \frac{e_b}{B} \quad \text{and} \quad 1 - \frac{e_a}{A} - \frac{e_b}{B}$$

that is,

$$1 \pm \left(\frac{e_a}{A} + \frac{e_b}{B}\right)$$

This relationship for relative error is the same as that for multiplication. The relationship for the division itself would be

$$\frac{A}{B}\left[1 \pm \left(\frac{e_a}{A} + \frac{e_b}{B}\right)\right]$$

It is easier to calculate errors in multiplication and division by using these equations for relative error, rather than that given for maximum error in multiplication.

By going through derivations such as those preceding, one can show that the maximum fraction of error in the result of a series of divisions or multiplications is the sum of the fractions of error in each number. For example, the maximum error limits in the calculation of AB/CD are

$$\frac{AB}{CD}\left[1 \pm \left(\frac{e_a}{A} + \frac{e_b}{B} + \frac{e_c}{C} + \frac{e_d}{D}\right)\right]$$

0.9-3 AVERAGE DEVIATION

Where many determinations of the same value are made, random errors (also called indeterminate errors) due to chance fluctuations should tend to average out. In such a case, the average value is considered to be better than any single determination. The average deviation, which is the average of the deviations of individual values from the average value, often is used as a measure of the error in these cases. Average deviation as a measure of indeterminate errors in repetitive measurements is described in Experiment 1.

Average deviation is the average of the sum of the absolute (disregarding sign) differences between observed values (X_i) and the average value (X_{ave}). The relation may be written as follows, where n is the number of observed values:

$$\frac{1}{n} \Sigma[(X_i - X_{ave})]$$

0.10 SLIDE RULE

A slide rule is very useful for carrying out the calculations in this laboratory manual and working chemistry problems in general. Multiplication and division can be learned in about five minutes. The accuracy of a 12-inch slide rule is about 1 part in 1000, per operation. This means that the slide rule can be read to 1 part in 1000. Of course, errors pile up when many numbers are multiplied or divided in succession. However, one percent accuracy is sufficient for most of the experiments in this book.

The slide rule (Figure 0-33) is essentially two logarithmic scales. Sections of the scale correspond to the logarithms of the numbers printed on the scale. Thus, if 2 is multiplied by 3, the length of scale D, corresponding to the logarithm of 2, is added to the length of scale C, corresponding to the logarithm of 3. The answer, read on the D scale, is 6.

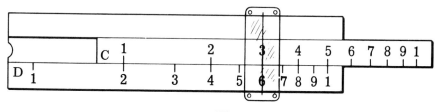

Figure 0-33. Slide rule (simplified).

You should try a more complicated problem with your slide rule. Let us multiply 23.2 by 10.8. Set the hairline slider at 232 on the D scale. Bring Point 1 of the C scale to coincide with the hairline. Then move the hairline to 108 on the C scale without moving the middle strip (C scale) of the slide rule. Read the answer where the hairline crosses the D scale. The answer is 2505. The decimal point must be obtained by approximating the answer. The answer is about 20×10, or 200, therefore, the answer must be 250.5. Notice that the answer is slightly wrong because of slide rule error.

If you wish to multiply 356 by 0.0402, set the hairline at 356 on the D scale. Then set the 1 line on the right-hand end of the C scale at the hairline, and move the hairline to 402 without moving the C scale. The answer on the D scale is 143, which must be 14.3, because $300 \times 0.04 = 12$. Notice that if we had used the 1 at the left end of the C scale, 402 would have been off the D scale.

Division is the reverse of multiplication. If you wish to divide 407 by 0.067, set the hairline at 407 on the D scale. Move the C scale so that 670 coincides with the hairline. Now move the hairline to 1 on the C scale (whichever end is on the D scale) and read 607 on the D scale. The answer is 6070, because 400/0.07 is about 6000.

Successive multiplications and divisions are also easy. For example, let us calculate $(760 \times 22.4 \times 4.13) \div (46 \times 273)$. Set the hairline at 760 on the D scale, and move the C scale so that 46 lines up with the hairline. (If we move the hairline to 1 on the C scale, we have divided by 46. But let us multiply directly by 22.4) Now move the hairline to coincide with 224 on the C scale without moving the C scale. [The result of $(760 \div 46) \times 22.4$ now can be read from the D scale.] To divide by 273, move the C scale so that 273 on the C scale coincides with the hairline. Move the hairline to 413 on the C scale, and read the final answer from the D scale, which is 560.

If you forget and are confused about how to use the slide rule, refresh your memory by doing a simple calculation to which you know the answer, like 2×2. Consult your slide rule book on how to use the logarithm scale or the sine scale. In these cases particularly, if you cannot remember which scale has the logarithm or sine, look for a number you know. For example, log 2 is 0.3, and the sine of 30° is 0.5.

Equipment and supplies

Two-pan analytical balance
Metal tag unknown

Time requirement

Two hours, if there are two students per balance

WEIGHING AN UNKNOWN WITH THE TWO-PAN ANALYTICAL BALANCE

The purpose of this experiment is to weigh an unknown object using the two-pan analytical balance. This assignment is superfluous if automatic balances are used in the course.

PROCEDURE

Study the discussion of the principles and operation of the two-pan analytical balance in Section 0.4-1 before coming to the laboratory class. The instructor will demonstrate the use of the balance. Obtain an unknown sample from the instructor, and record the number of this unknown in your notebook. Weigh the unknown four or five times. Handle the unknown with forceps if the weighing is to be to tenths of a milligram. If the weighing is to the nearest milligram, the moisture from your fingers will not affect the weighing significantly.

Average your weighings. Estimate the error in the average weight as the average deviation (see the next section). Include in your report, in addition to the weights of the unknown and the average weight with the average deviation, sample data and calculations showing exactly how you found the sensitivity of the balance. Estimate the readability of your analytical balance (perhaps 0.1 mg). Calculate the maximum error in the weight of the unknown (see Section 0.9) due to the error in reading the balance and the errors in the set of weights. Compare this maximum error with the average deviation, and discuss the difference, if any.

AVERAGE DEVIATION

The average deviation is the average of the deviations of individual weighings from the average weight. For example, consider the following four weighings:

Weight	Deviation
2.1045 g	0.0002 g
2.1039	0.0004
2.1046	0.0003
2.1042	0.0001
4 ⌟8.4172	4 ⌟0.0010
2.1043 g Average	0.0003 g Average

The average weight is 2.1043 g. The deviation of the first weighing from the average weight is 2.1045 g — 2.1043 g = 0.0002 g. The average of all four deviations is 0.0003 g.

This average deviation is a measure of the random errors (indeterminate error) in the weighings. Consequently, the result may be reported as 2.1043 g ± 0.0003 g. This is a common method of estimating indeterminate errors in repetitious measurements.

EXPERIMENT 2

Equipment and supplies

Analytical balance
Two 50-ml beakers
Two watch glasses
Drying oven

2 g Barium chloride dihydrate (duplicate samples)

Time requirement

Two and one half hours

GRAVIMETRIC DETERMINATION OF WATER

The purpose of this experiment is to determine the percentage of water by weight in a hydrated salt.

INTRODUCTION

The method to be used involves weighing a sample, removing the water, and then reweighing the sample. The difference in weight is the weight of water lost. This method often is called the indirect method, because the water lost is not weighed by itself. In the direct method, the water is driven off, collected, and weighed.

The conditions needed to remove water from a particular substance depend on the properties of the substance. Many hydrated salts lose water of hydration easily at temperatures above the boiling point of water. Such is the case with barium chloride dihydrate ($BaCl_2 \cdot 2H_2O$) in this experiment. The oven temperature specified in the experiment is 150°C, rather than 105°C to 110°C, because the oven is opened so often in a large class that the temperature is, on the average, much lower. Many hydrated salts require higher temperatures to drive off all the water. The dehydrated salts are often hygroscopic (they absorb water), so they must be kept covered after drying to prevent absorption of atmospheric moisture. Compounds that are not stable when heated may be dried by subjecting them to vacuum, or by exposing them to the dry atmosphere of a desiccator containing a strong dehydrating agent such as phosphorus pentoxide or concentrated sulfuric acid.

PROCEDURE

If this is your first experiment with the analytical balance, study the discussion of the principles and operation of the analytical balance in Section 0.4 before coming to the laboratory class. The instructor will demonstrate the use of the balance.

If equipment permits, the following analysis should be done with duplicate samples in two beakers. In general, the average result of two

Figure 2-1. Covered beaker with weighing bottle.

analyses is more accurate than for a single sample. And if you should drop one of the two beakers, the analysis is not ruined completely. So if possible, do determinations on duplicate samples at the same time, but be sure that the beakers are marked.

Clean a 50-ml beaker. Heat it in an oven at 150°C. (If the oven is so dirty that dust and dirt might fall into the beaker, the oven should be cleaned. However, the beaker can be protected from such contamination by covering it with a watch glass.) When the beaker is dry, take it from the oven, using tongs or a towel. Let it cool to room temperature (8-10 minutes). When it has cooled, weigh the beaker on the analytical balance accurately. Never attempt to weigh a hot object on the balance.

Weigh out about 1 g of barium chloride dihydrate (or other hydrated salt) on a piece of smooth paper, using a rough balance. Add this salt to the accurately weighed beaker and reweigh accurately on the analytical balance.

Now heat the sample and the beaker in the drying oven for half an hour at 150°C (or longer at 110°C). (If contamination of your sample in the oven is possible, the beaker should be covered with a watch glass that is supported by three glass hooks, as shown in Figure 2-1. These glass hooks are made from glass rod and allow freer circulation of air during drying.) After half an hour, take the beaker out of the oven and cover it with a watch glass to minimize access to atmospheric moisture. After the beaker has cooled, weigh it without the watch glass on the analytical balance. Heat the sample in the beaker again for half an hour at 150°C. Take it out and let it cool as before. Reweigh it accurately. If the weight is not the same this time as it was after the first heating (within your weighing error), repeat the drying procedure until constant weight is attained.

Compute the percentage by weight of water in the hydrated salt. If the salt is barium chloride, show in your report how the formula of the hydrated salt could be calculated from this value.

For very accurate work where weighings to 0.1 mg are desired, a glass-stoppered weighing bottle should be used (Figure 2-2) instead of a beaker. Furthermore, the weighing bottle should not be touched with the fingers. A piece of paper may be folded once or twice to make a strip, and the bottle handled with this strip of folded paper, as in Figure 2-2. The bottle should be heated with the glass stopper ajar inside a beaker, as shown in Figure 2-1. Also, for accurate work the weighing bottle should be cooled in a desiccator with the glass stopper of the weighing bottle ajar.

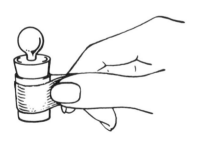

Figure 2-2. Handling the weighing bottle.

SUPPLEMENTARY READING

Hammond, Osteryoung, Crawford, and Gray (Chapter 1, Section 2)

QUESTIONS

1. If the sample of barium chloride dihydrate is *not* heated repeatedly until constant weight is attained, what difference is this likely to make in the percentage of water determined?

2. If a large piece of dirt weighing 10 mg falls into the beaker *after* the sample is weighed before heating, what effect would this have on the percentage of water found?

3. In what sense is the single-pan automatic balance (or two-pan analytical balance) a "balance"? What is being balanced against what?

PROBLEMS

1. Calculate the maximum indeterminate error in the percentage of water found for the hydrated salt (see Section 0.9). Assume a weighing error of one half the smallest weight read. If your weighings were to 1 mg, the error can be assumed to be ± 0.5 mg in each weighing.

2. How could the percentage of indeterminate error be reduced in this experiment, still using the same equipment?

3. A 1.00-g sample of blue hydrated copper sulfate is heated at 200°C to constant weight. The weight after drying is 0.639 g. Calculate the number of water molecules in the formula $CuSO_4 \cdot xH_2O$ and the percentage by weight of water of hydration.

EXPERIMENT 3

Equipment and supplies (for two determinations)

Analytical balance that can weigh to 1 mg or better

Two evaporating dishes, or two 100-ml beakers

Two watch glasses

Drying oven

Two 100-ml water samples

Time requirement

Two hours total, depending on access to the balance. Forty minutes to boil down the 100 ml of water.

GRAVIMETRIC DETERMINATION OF TOTAL RESIDUE OF DISSOLVED SOLIDS IN WATER

The purpose of this experiment is to determine the weight of dissolved solids in a sample of water.

INTRODUCTION

Water as it occurs in nature contains various amounts of dissolved substances, depending on its origin; for example, rain, lakes, rivers, wells, or oceans. The dissolved substances may be gases, solids, and, less commonly, liquids. Undesirable gases are removed easily from the water by aeration, but dissolved solids are not removed easily. These dissolved substances affect the suitability of the water for drinking or other uses. Among the standards set by the Federal Government for drinking water is that there cannot be more than 500 mg of dissolved solids per liter (500 parts per million).

An example of highly mineralized water is found in the Santa Barbara, California, area. The amount of dissolved solids varies during the year, but is almost always more (0.5-1.2 g liter^{-1}) than the recommended standard. The dissolved materials in the water seem to have no adverse effect on health. The water contains no transition or heavy metals, because the water is slightly basic. Also, dissolved fluoride ion, which is naturally present, is thought to be beneficial in protecting the teeth of children from decay. The water is said to satisfy requirements for bacterial content even before chlorination (usually with sodium hypochlorite). On the negative side, sometimes there is a slightly unpleasant odor in the water that is caused by minute quantities of sulfur compounds originating from sulfur springs. A substantial concentration of magnesium sulfate (epsom salts) in the water has a cathartic effect on some newly arrived visitors to the area, although the effect disappears in a day or two. Nevertheless, the principal objection to the dissolved solids is the scale deposited by the water in taps, water heaters, boilers, and plumbing fixtures, and the problems of laundering in hard water (water containing large amounts of divalent cations like Ca^{2+}, Fe^{2+}, and Mg^{2+}).

The following experiment is suitable for a first assignment using the analytical balance. Students may wish to analyze water other than the local tap water. If they do, they should be sure to indicate the source of the water being analyzed. The water samples may be analyzed qualitatively for ionic composition, using the procedures of Experiment 26, Part 7. Chloride ion concentration can be determined quantitatively using the procedure of Experiment 5, Part C. The acidity or basicity of water can be measured by the method of Experiment 23.

PROCEDURE

If this is your first experiment with the analytical balance, study the discussion of the principles and operation of the analytical balance in Section 0.4 before coming to the laboratory class. The instructor will demonstrate the use of the balance.

If equipment permits, the following analysis should be done with duplicate samples. In general, the average of two results is more accurate than a single result, and spoiling one sample does not ruin the analysis. So, if possible, do duplicate runs, but make sure that your evaporating dishes or beakers are marked. Your initials and "1" and "2" on the etched surface in pencil works well. An evaporating dish is recommended in this experiment because the final slow boiling or evaporation is faster in an evaporating dish. However, beakers can be used.

Clean an evaporating dish or a 100-ml beaker. Heat it in an oven at 150°C. (Actually, only a temperature of 105°C-110°C is needed, but because in a large laboratory class the oven inevitably is opened often, the temperature is best set at 150°C.) If the oven is so dirty that dust and dirt might fall into the beaker, the oven should be cleaned. However, the dish or beaker can be protected from such contamination by covering it with a watch glass in the manner shown in Figure 2-1. When the dish or beaker is dry (10-15 minutes), take it from the oven using tongs or a towel; 105°C-150°C is hot! Let it cool to room temperature on an asbestos square or wire gauze (8-10 minutes). When it has cooled, weigh the dish (or beaker) on the analytical balance accurately. Never attempt to weigh a hot object on the balance.

Place the evaporating dish (or beaker) on a wire gauze supported on a ring stand (Figure 0-21). The wire gauze should be about four inches above the burner top. Obtain 100 ml of water sample (tap water or other source) in a graduated cylinder. The weight of most samples of water that have only small amounts of dissolved solids can be assumed to be 1.0 g per 1 ml for a rough determination. If the determination is to be more accurate, say ±2% or less, the water sample should be weighed. To do so, weigh the graduated cylinder without the water, and then with the 100 ml of water sample, on a triple beam balance to 0.1 gram. Add about 30 ml of water to the dish or beaker, and heat the dish or beaker with a burner (Figure 0-21) so that the water boils gently. The rate of boiling can be controlled by moving the burner away from the center toward the side of the dish. Add more of the water sample when the level drops. When the volume of water sample has finally boiled down to about 10-20 ml, it may be advisable to cover the dish with a watch glass to catch spattering from the concentrated solution. When the dish is dry, rinse spatterings off the watch glass into the dish with a little distilled water from a plastic squeeze bottle. Because the watch glass is hot, it will be necessary to hold it with tongs. Evaporate the small amount of water in the dish to dryness by heating the dish gently without the watch glass. Note and record the color of the residue.

Dry the dish or beaker in the oven for 10-15 minutes. Let the dish or beaker cool on an asbestos square or wire gauze for 8-10 minutes. Then weigh it accurately on the analytical balance. The difference in weight between this and the first weighing of the dish is the weight of dissolved solids in the water sample. Express the total residue in mg liter^{-1}, or parts per million (abbreviated ppm). Compare the result to the Government standard.

SUPPLEMENTARY READING

Hammond, Osteryoung, Crawford, and Gray (Chapter 1, Section 2)

QUESTIONS

1. If the residue is not heated to dryness but is still somewhat moist, what difference is this likely to make in the amount of dissolved solids found?

2. If a large piece of dirt, say 10 mg, falls into the dish just before the sample is weighed the second time, what effect would this have on the weight of dissolved solids found?

3. In what sense is the single-pan automatic balance (or two-pan analytical balance) a "balance"? What is being balanced against what?

PROBLEM

1. Calculate the maximum indeterminate error in the total residue found (see Section 0.9). Assume that measurements are ± one half the smallest unit of the measurement.

EXPERIMENT 4

Equipment and supplies

Analytical balance

Two 250-ml beakers

Burner, wire gauze, ring stand

Hood

Plastic squeeze bottle

Drying oven

20 ml 6M Nitric acid

25 ml 3M Sulfuric acid

1 ml 6M Hydrochloric acid

Two 8-inch lengths of copper wire, B & S No. 14 or 16, bare

Two 1.5-g samples of silver-copper alloy for duplicate samples (brazing wires can be used as unknowns)

Time requirement

Two hours

ANALYSIS OF SILVER-COPPER ALLOY

The purpose of this assignment is to determine the percentage of pure silver in an alloy of silver and copper.

PROCEDURE

Clean a 250-ml beaker and rinse it with distilled water. Dry the beaker in a drying oven at 105°C-150°C. Let it cool, and weigh it on an analytical balance. Put the sample (about 1 g) of alloy in the beaker, and reweigh it on the analytical balance.

Add 10 ml of $6M$ nitric acid to the beaker. In dilute nitric acid, the reaction may be described as

$$4H^+ + NO_3^- + 3Ag \rightarrow 3Ag^+ + 2H_2O + NO$$

and in more concentrated nitric acid

$$2H^+ + NO_3^- + Ag \rightarrow Ag^+ + H_2O + NO_2$$

Under the conditions of this experiment, both nitric oxide and nitrogen dioxide (brown fumes) will be formed. You may need to warm the solution slightly to start the reaction. Place the beaker on a wire gauze supported on a ring stand. Heat the beaker gently with the burner flame. The wire gauze spreads the heat out, and makes it more uniform. Avoid overheating and rapid reaction, because material would be lost by spattering.

After the metal is dissolved completely, cool the solution and add 25 ml of $3M$ sulfuric acid to it. Place the beaker on a wire gauze supported on a ring stand in the hood. With the smallest possible flame, evaporate the solution in the hood (so that noxious vapors do not pollute the laboratory) until all of the nitric acid has been removed. This point will be indicated by evolution of dense, white, choking fumes of sulfur trioxide, SO_3. Why must the nitric acid be removed at this point in this procedure?

Let the beaker cool. It now contains concentrated sulfuric acid that reacts vigorously with water. So very carefully add, in small portions at first, a total of 125 ml of distilled water. Swirl to mix after each addition. A white precipitate of silver sulfate (Ag_2SO_4) probably will form, but this will dissolve upon heating. Heat the solution until it is quite hot, but do not allow it to boil. Stir with a clean glass stirring rod until a clear solution is obtained.

Obtain about an 8-inch length of copper wire (B & S gauge No. 14) and coil it so that only a short length will protrude as a handle from the beaker. Place this coil of wire in the hot solution. You may enjoy watching the formation of silver crystals for five minutes or so. The reaction in which the copper metal dissolves and the silver ions in solution form metallic silver may be written

$$Cu + 2Ag^+ \rightarrow 2Ag + Cu^{2+}$$

When you have satisfied your curiosity, gently stir, and shake the wire to free the copper surface so further reaction can occur.

When no more silver appears in five minutes after the wire has been shaken, withdraw about 1 ml of solution with a clean medicine dropper into a clean test tube, both of which have been rinsed with distilled water. Add a few drops of $6M$ hydrochloric acid to the test tube, and look for the presence of a milky precipitate, which is silver chloride, an insoluble salt. Silver chloride is so insoluble that the precipitate will form even if only a very small amount of silver ion is present. The reaction may be written

$$Ag^+ + Cl^- \rightarrow AgCl\downarrow$$

If the test is positive, continue to shake the copper wire until a negative test for silver ion is obtained. It may be necessary to warm the solution again.

After the reaction is complete, let the silver metal settle to the bottom of the beaker, and carefully decant (pour off) as much of the solution as possible without loss of silver crystals. Rinse the silver and the beaker, by directing a stream of distilled water from a squeeze bottle at the inside walls of the beaker. Then decant the wash water, taking care not to lose any silver. It may be necessary to rinse the outside of the beaker also. Rinse the silver and the inside of the beaker twice more. Dry the beaker containing the silver in a drying oven to constant weight. Weigh the beaker and silver after it has cooled. If a drying oven is not available, the beaker can be dried by gently heating it with a burner.

Compute the percentage by weight of silver in the alloy.

SUPPLEMENTARY READING

Hammond, Osteryoung, Crawford, and Gray (Chapter 1, Section 2)

QUESTIONS

1. Would the percentage of silver found be larger, or smaller, than the true value if the silver crystals were not completely dry when weighed?

2. What effect would incomplete precipitation of the silver metal by copper have on the results?

3. Why must the nitric acid be removed before the precipitation of silver with the copper wire?

PROBLEM

The reaction used in this experiment was one of those used by T. W. Richards (Harvard University) to determine the atomic weight of copper (1886-1891). Suppose that an excess of cold silver nitrate solution dissolved 35.788 g of pure copper to make 60.703 g of pure silver. Show how the atomic weight of copper can be obtained from these values, if the value for silver is known to be 107.88 g per gram-atom.

EXPERIMENT 5

Equipment and supplies

Procedure A (for one determination)

One 70-ml evaporating dish
Burner, wire gauze, ring stand
One watch glass
One 250-ml beaker
One plastic squeeze bottle
Analytical balance

1 g Silver, pure
7 ml 6M Nitric acid, CP
3 ml Concentrated hydrochloric acid
Boiling stone

Procedure B (for one determination)

One beaker, 100- or 150-ml
One watch glass
Burner, wire gauze, ring stand
One plastic squeeze bottle
One stirring rod
One filter crucible (sintered glass or equivalent; see procedures—to be approved by instructor for one period)
Drying oven
One rubber policeman
Analytical balance

1 g Silver, pure (0.5-g sample or less may be used)
7 ml 6M Nitric acid, CP
5 ml 6M Hydrochloric acid
10 ml 15M Ammonia (for cleaning filter)

Procedure C (for one determination)

One 250-ml beaker
Burner, wire gauze, ring stand
One stirring rod
One filter crucible, sintered glass
One filter flask, rubber ring
One rubber policeman
Oven
Analytical balance

Unknown sample
1 ml 6M Nitric acid
1 Blue litmus paper
10 ml 1M Silver nitrate

Time requirement

Two hours for each part, depending on the availability of balances

THE ATOMIC WEIGHT OF CHLORINE, AND THE GRAVIMETRIC ANALYSIS OF SILVER OR CHLORINE AS SILVER CHLORIDE

In this experiment, silver is dissolved in nitric acid and converted to silver chloride. From the weight relationship, the atomic weight of chlorine is determined. In Part C, the chlorine content of an unknown is determined.

HISTORICAL NOTE

In addition to the utility of an accurate set of combining weights, the precise measurement of atomic weights received incentive from two other sources. These sources, Prout's hypothesis, and Mendelyeev's periodic chart, were the result of the desire to organize the 60 or so known elements into a rational system.

In 1815, Prout* suggested that the elements were composed of a fundamental particle called the "protyle," which was thought to constitute the simplest atom, hydrogen. All other elements consisted of multiples of the fundamental particle, and, consequently, the atomic weights of the elements must be exact multiples of the atomic weight of hydrogen. This theory seems fairly good if the first 16 elements are considered, since the atomic weights approximate whole numbers. However, Berzelius'[2] atomic weight for chlorine was 35.4 (on a scale where O = 16), clearly violating the theory of Prout. It was suggested that the experimental value of 35.4 was not a whole number because of experimental error. To test the theory, more accurate determinations of the atomic weight of chlorine were necessary. Very accurate measurements were carried out by Stas,[3] Marignac,[4] and Richards.[5]

One procedure, yielding the atomic weights of chlorine, potassium, and silver, relative to oxygen, involved the decomposition of potassium chlorate to potassium chloride and oxygen, the formation of silver chloride from potassium chloride, and the formation of silver chloride from silver. The silver chloride procedure was originated by Berzelius, but also was used by the later workers.

Stas, for example, heated 127.2125 g of potassium chlorate. He obtained 77.4023 g of potassium chloride as a residue and found from

---

*Superior numbers refer to the reference books listed in Appendix 1.

the difference that 49.8102 g of oxygen evolved

$$KClO_3 \rightarrow KCl + \frac{3}{2}O_2 \uparrow$$

Because three oxygen atoms were lost for each chlorate, the following proportion can be written, which yields the molecular weight of potassium chloride (mol wt$_{KCl}$) based on oxygen as 16.00

$$3 \times 16 : \text{mol wt}_{KCl} = 49.8102 : 77.4023$$

or

$$\frac{\text{mol wt}_{KCl}}{3 \times 16} = \frac{77.4023}{49.8102}$$

$$\text{mol wt}_{KCl} = 74.59 \text{ g mole}^{-1}$$

In order to get the atomic weights of potassium and chlorine from this value, more information is needed. Marignac was able to obtain 27.732 g of silver chloride from 14.427 g of potassium chloride by adding an excess of silver nitrate solution to a solution containing the potassium chloride

$$AgNO_3 + KCl \rightarrow AgCl \downarrow + KNO_3$$

From these weights, the ratio of the molecular weight of KCl to that of AgCl can be obtained, and the molecular weight of silver chloride calculated

$$\frac{\text{mol wt}_{KCl}}{\text{mol wt}_{AgCl}} = \frac{74.59}{\text{mol wt}_{AgCl}} = \frac{14.427}{27.732}$$

$$\text{mol wt}_{AgCl} = 143.37 \text{ g mole}^{-1}$$

If the combining weight of silver in a certain amount of silver chloride can be determined, the atomic weights of silver, chlorine, and potassium can be obtained. The ratio of silver to silver chloride can be obtained by dissolving the silver in nitric acid

$$3Ag + 4HNO_3 \rightarrow 3AgNO_3 + 2H_2O + NO \uparrow$$

and adding excess hydrochloric acid to form silver chloride, as in your laboratory assignment

$$AgNO_3 + HCl \rightarrow AgCl \downarrow + HNO_3$$

The conversion of silver to silver chloride also can be achieved by heating silver in a stream of chlorine gas

$$Ag + \frac{1}{2}Cl_2 \rightarrow AgCl$$

As an example, Stas, after heating 101.519 g of silver in chlorine, obtained 134.861 g of silver chloride. Using the molecular weight of silver chloride calculated previously, we find

$$\frac{\text{at. wt}_{Ag}}{\text{mol wt}_{AgCl}} = \frac{\text{at. wt}_{Ag}}{143.37} = \frac{101.519}{134.861}$$

$$\text{at. wt}_{Ag} = 107.93 \text{ g mole}^{-1}$$

$$\text{at. wt}_{Cl} = 143.37 - 107.93 = 35.44 \text{ g mole}^{-1}$$

$$\text{at. wt}_{K} = 74.59 - 35.44 = 39.15 \text{ g mole}^{-1}$$

Revision and remeasurement of atomic weights in the latter part of the nineteenth century also were stimulated by the development of the periodic law. Although relationships had been noticed earlier between certain atomic weights, in 1864, Newlands first arranged the elements in a series by their atomic weights. It appeared that similar elements were found at every eighth one (the rare gases were not discovered until after 1900). For example, in the following list Li, Na, and K are similar

H, Li, Be, B, C, N, O, F, Na, Mg, Al, Si, P, S, Cl, K

This he called the law of octaves. This law was ridiculed for its suggestion of musical scales.

Lothar Meyer and Dmitri Mendelyeev independently showed that "the properties of the elements are periodic functions of their atomic weights." The graph of atomic weight versus atomic volume, which shows this periodicity beautifully, was published by Meyer in 1870.[6] Mendelyeev[7] actually published a table of the elements first. The elements were arranged in increasing order of atomic weights vertically, and elements of similar properties were arranged side by side horizontally. From vacant spots in the table he was able to predict the properties of then unknown elements such as germanium, gallium, and scandium. However, it became apparent that some atomic weights were not accurate, and others appeared to be incorrect. For example, iodine and tellurium were minor exceptions to the periodic law as stated. To settle this problem, more accurate atomic weights were needed.

PROCEDURE A: ATOMIC WEIGHT BY THE DRY METHOD

Clean and dry a 70-ml evaporating dish or casserole. Weigh the dish accurately. Record the weight of the dish. Place a piece of silver weighing about one gram in the dish. Weigh the dish and piece of silver. Record the weight.

Place the dish (or casserole) on a wire gauze supported on a ring stand. To the dish, add about 7 ml of dilute (6M) nitric acid, CP (chemically pure). Cover the dish with a watch glass, convex side down, to avoid loss by spattering. To start and maintain the reaction, the dish may be heated gently with a Bunsen burner. The nitric acid dissolves the silver metal

$$3Ag + 4H^+ + NO_3^- \rightarrow 3Ag^+ + 2H_2O + NO$$

When the reaction is complete, that is, when the silver has disappeared completely, remove the dish from the wire gauze. Place a 250-ml beaker about two-thirds full of water on the wire gauze. Add a boiling stone. Set the dish (or casserole) containing the silver solution on the beaker and heat the water to boiling. Remove the watch glass, carefully letting the drops adhering to the convex side drip into the dish. Rinse the watch glass with a stream of water from a plastic squeeze bottle. Do not use more water than necessary. Add about 2 ml of concentrated hydrochloric acid to the dish. A white precipitate of silver chloride will form

$$Ag^+ + Cl^- \rightarrow AgCl\downarrow$$

All the substances present in the dish are volatile except the silver chloride. That is, they may be removed by gentle heating, leaving the pure, dry silver chloride. Boil the water in the beaker. The contents of the dish will evaporate slowly without bumping. When the contents of the dish seem to be dry, moisten them with about 1 ml of concentrated hydrochloric acid, and dry once more. To prevent noxious vapors from polluting the laboratory air, perform all such operations under a hood.

When the dish is dry again, remove it from the beaker, and remove the beaker from the wire gauze. Set the dish on the wire gauze, and heat it gently with a very small flame for several minutes. Avoid heating the dish too strongly because the contents may spatter. Nevertheless, heat the dish strongly enough so that it is completely dry, without melting the silver chloride. When the dish has cooled to room temperature, weight it again. Record the weight.

If the atomic weight of silver is known, that is, 107.9, and the formula of silver chloride is AgCl, find the atomic weight of chlorine from your data.

PROCEDURE B: ATOMIC WEIGHT BY THE WET METHOD

Clean a 100- or 150-ml beaker. Rinse it three times with distilled water. Weigh a piece of silver (approximately one gram) on an analytical balance. Record the weight. Place the silver in the beaker. To the beaker, add about 7 ml of dilute ($6M$) nitric acid, CP (chemically pure). Cover the beaker with a watch glass, convex side down, to avoid loss by spattering. Place the beaker on a wire gauze supported on a ring stand, and warm it gently with a Bunsen burner.

When the reaction

$$3Ag + 4H^+ + NO_3^- \rightarrow 3Ag^+ + 2H_2O + NO$$

is complete, that is, when the silver has completely disappeared, turn off the burner and rinse the bottom of the watch glass with a stream of distilled water from a squeeze bottle. In doing so, add about 20 ml of distilled water to the solution. Let it cool to near room temperature.

While stirring with a glass stirring rod, slowly add slightly more hydrochloric acid than is needed to precipitate all the silver. This should require about 11-15 ml of $1.0M$ hydrochloric acid. This $1.0M$ HCl can be made by mixing 10 ml of $6M$ HCl (the dilute HCl on the reagent shelf) with 50 ml of distilled water. To test for complete precipitation, let the white precipitate settle, then add a little more HCl.

For best results, it is a good idea to carry out the precipitation and later procedures in subdued daylight, because silver chloride is decomposed slowly by light

$$2AgCl \rightarrow 2Ag + Cl_2$$

A large excess of chloride ion must be avoided because of formation of soluble silver chloride complex, which would reduce the weight of silver chloride precipitate

$$AgCl + Cl^- \rightarrow AgCl_2^-$$

The precipitation is done slowly with stirring to avoid coprecipitation of ions other than silver. Also, the precipitate coagulates better when formed slowly, and will not adhere to the sides of the beaker. A heavily scratched beaker should not be used in this experiment because the silver chloride may adhere to the scratched glass. If the precipitate does not coagulate well, it may be desirable to heat the beaker nearly to boiling, stir, and allow to cool.

While the silver is dissolving, clean (see below), rinse with distilled water, and suck a sintered glass filter crucible dry on the filter flask. Place the filter crucible in a beaker (cover with a watch glass if the oven is dirty), and dry it in the drying oven. Let it cool and weigh the filter crucible on an analytical balance. Touch the crucible with dry fingers only. Since weighings will be to the nearest milligram, the change in weight due to sweat on the fingers will not be important.

After precipitation of the silver chloride, as described previously, filter the silver chloride into the filter crucible, using the filter flask and suction. Carefully rinse any remaining silver chloride from the beaker onto the filter with a stream of distilled water from a squeeze bottle. A rubber policeman may be needed in extreme cases to get the precipitate onto the filter. Consult Section 0.8 on filtration before carrying out the filtration. Rinse the precipitate with a small amount of water from the squeeze bottle. Suck all the water off the precipitate. Then place the crucible in the beaker, and dry in the drying oven at $110°C$ for an hour. Let cool, and weigh the crucible. Heat the crucible again, let it cool, and reweigh. This is done to check that the precipitate is really dry.

A glass fiber filter on a porcelain Gooch crucible (with holes in the bottom) is cheaper and as good for this experiment. The fiber filters may be purchased from H. Reeve Angel & Co., Clifton, New Jersey.

If the atomic weight of silver is known to be 107.9, and the formula of silver chloride is AgCl, calculate the atomic weight of chlorine from your data.

The filter crucible should be cleaned by washing it with ammonia. Silver chloride is soluble in ammonia

$$AgCl + 2NH_3 \rightarrow Ag(NH_3)_2^+ + Cl^-$$

Return the clean filter crucible to the storeroom.

PROCEDURE C: DETERMINATION OF CHLORIDE IN A WATER-SOLUBLE CHLORIDE OR AQUEOUS SOLUTION

Weigh a 250-ml beaker on a triple beam balance (not an analytical balance) as accurately as possible (±0.1 g). Place 100 ml of an aqueous chloride solution (or a volume containing at least 0.1 g of chloride if possible) in the 250-ml beaker and reweigh on the triple beam balance. The difference is the weight of the sample. A sample of tap water, or other water, can be analyzed by this method. Because the amount of dissolved material in tap water is small, a larger volume of the water should be concentrated to 100 ml by boiling. This will give a more accurate determination. Incidentally, bromide and iodide will precipitate also if present, whereas sulfate and carbonate will not.

If the sample to be analyzed is a solid, weigh a 250-ml beaker accurately on an analytical balance. Add about 0.5 g of the sample and reweigh the beaker and sample on the analytical balance. The difference in weights is the weight of the sample. Add 100 ml of distilled water to the beaker and stir to dissolve.

Acidify the beaker of chloride solution (either the aqueous solution or that from the solid sample) by adding 1 ml of $6M$ nitric acid. Check the solution with blue litmus paper to make sure that it is acidic. (Blue litmus paper turns red in acidic solution.) Now add $1M$ silver nitrate solution slowly while stirring. A white precipitate of silver chloride forms

$$Ag^+ + Cl^- \rightarrow AgCl\downarrow$$

An excess of silver nitrate is to be added. If the approximate amount of chloride present is known, the approximate amount of silver nitrate required can be calculated. (If the solid sample were sodium chloride, 8.5 ml of $1M$ silver nitrate would be needed for a 0.5-g sample.) If the approximate amount of chloride is not known, add 3 ml of $1M$ silver nitrate, stir, and let the precipitate settle. Then add a few drops more of the silver nitrate, and watch for the formation of more white precipitate. If more silver chloride forms, the chloride has not been all precipitated. So add more silver nitrate solution. Repeat until no more silver chloride forms. After the chloride has been precipitated as silver chloride, heat the solution and precipitate just to boiling, stir, and let cool. This helps to coagulate the precipitate, thus makes filtering easier.

For best results it is a good idea to carry out the precipitation and later procedures in subdued daylight, because silver chloride is decomposed slowly by light

$$2AgCl \rightarrow 2Ag + Cl_2$$

The precipitation is done slowly with stirring to avoid coprecipitation of ions other than silver. Also, the precipitate coagulates better and will not adhere to the sides of the beaker. A heavily scratched beaker should not be used in this experiment because silver chloride may adhere to the scratched glass.

Clean a filter crucible. Rinse it with distilled water and suck it dry on the filter flask (see Section 0.8 and Figure 0-29). Place the filter crucible in a beaker (covered with a watch glass if the oven is dirty), and dry it in the oven. Let cool, and weigh the filter crucible on an analytical balance. If weighings are to be done to 1 mg, one can touch the crucible with dry fingers. If weighings are to be to 0.1 mg, the crucible should be handled with tongs or a small piece of paper as shown in Figure 2-2.

After precipitation of the silver chloride, as described previously, filter the silver chloride into the filter crucible, using the filter flask and suction. Carefully rinse any remaining silver chloride from the beaker onto the filter with a stream of distilled water from a squeeze bottle. A rubber policeman may be needed in extreme cases to get the precipitate onto the filter. Consult Section 0.8 on filtration before carrying out the filtration. Rinse the precipitate with a small amount of distilled water from the squeeze bottle. Suck all the water off the precipitate. Then place the filter crucible in a beaker, and dry it in the oven at 110°C for one half hour. Let cool, and reweigh. This is done to be sure that the precipitate is really dry.

Calculate the percent of chlorine by weight in the sample. The atomic weights of chlorine and silver are known for this calculation.

The filter crucible should be cleaned by washing it with ammonia. Ammonia has a very pungent odor. Do not drop the crucible when you pour in some ammonia. Silver chloride is soluble in ammonia

$$AgCl + 2NH_3 \rightarrow Ag(NH_3)_2^+ + Cl^-$$

Return the clean filter crucible to the storeroom.

A word of caution about silver nitrate is in order. Silver salts in general are light sensitive, and on skin and clothing they eventually will turn black. Thus you should wash off immediately any silver salts on your skin or clothing. Dilute ammonia will dissolve solid silver salts.

SUPPLEMENTARY READING

Hammond, Osteryoung, Crawford, and Gray (Chapter 1, Section 8)

QUESTIONS

1. In Procedure A, how will the results be affected if the silver chloride is heated to the melting point to remove the last traces of water? When silver chloride is dried at 260°C, 0.01% of water remains. However, silver chloride evaporates readily at the melting point, about 445°C, but only slowly below the melting point.

54 The atomic weight of chlorine

2. In Procedure B, what will be the effect on the results if the silver chloride precipitate on the filter is washed with much water? The solubility of silver chloride is very small, about 1.4 mg per liter at room temperature.

3. If the precipitation of silver chloride is carried out rapidly, other metal ions in the solution (if there are any) may be carried down into the solid also (coprecipitation). How would this affect results?

4. Suggest a method for determination of the atomic weight of bromine (Br).

PROBLEMS

1. Estimate your error quantitatively (maximum error), assuming that the errors involved are only in the weighings.

2. About 1811-1812,[2] Berzelius stated that he dissolved 3.00 g of pure silver with nitric acid, added pure muriatic (hydrochloric) acid, and finally fused the product in the flask. The colorless horn silver (silver chloride) left weighed 3.98 g. If the formula weight of silver chloride is known to be 143.4, calculate the atomic weight of silver and of chlorine.

3. Using Procedure C, a student dissolved a 1.000-g sample of an impure chloride salt, acidified the solution, and treated it with a slight excess of silver nitrate solution. The dried precipitate of silver chloride weighs 2.000 g. Calculate the percentage of chlorine by weight in the chloride sample.

EXPERIMENT 6

Equipment and supplies

Clock with second hand

One 250-ml beaker

One 400-ml beaker

One 600-ml beaker

One square piece of cardboard

One special 0.1°C thermometer, −1°C-50°C

One section of rubber tubing (see procedure)

Platform balance

Burner, wire gauze, ring stand

Tongs

Paper towels

Ice

Piece of metal (to fit 250-ml beaker)

Time requirement

Two hours

HEAT CAPACITY AND HEAT OF FUSION

The purpose of this experiment is to determine the heat of fusion of ice and the heat capacity of a metal. From the semiempirical rule of Dulong and Petit, the atomic weight of the metal will be estimated. (See Supplementary Reading list at end of this experiment.)

HISTORICAL NOTE

In 1723, Brooke Taylor[8] found that the rise in temperature upon addition of hot water to cold water depended directly on the relative amount of hot water added. About 1760, Joseph Black used a hollowed-out block of ice as a calorimeter. The amount of heat added by a hot object was proportional to the weight of melted ice poured out afterward. In his lectures published in 1803, he distinguished between the intensive quantity, temperature, and the extensive quantity, the amount of heat. He also originated the idea of specific heat, after noticing that not all materials at the same temperature contained the same amount of heat. In 1780, Lavoisier and Laplace reported the use of an iron apparatus (and invented the word *calorimeter* for it) in which heat was measured by the weight of ice melted. They obtained 75 cal g^{-1} as the heat of fusion of ice, and also determined specific heats and heats of reaction. Dulong and Petit[9] noted the empirical fact that all solid elements, especially metals, had the same "atomic heat" of about 6 cal deg^{-1} $mole^{-1}$. This atomic heat is the heat capacity of a mole of a metal, which is the specific heat (calories per degree per gram) multiplied by the atomic weight. In other words, the law of Dulong and Petit states that equal numbers of metal atoms in the solid state have the same heat capacity. This rule was very useful for determining approximate atomic weights.

INTRODUCTION

When an object absorbs heat without undergoing a phase transition, its temperature always rises. The amount of heat absorbed is usually closely proportional to the temperature rise; in other words, the ratio

58 Heat capacity and heat of fusion

of heat absorbed to temperature rise is nearly constant for a given object. This ratio is called the *heat capacity*. The obvious statement, that for the same temperature change the heat capacity of two objects taken together must be exactly the sum of their individual heat capacities, leads to the definition of *specific heat capacity* as the heat capacity of one gram of a pure substance (in units of calories per degree per gram). The heat capacity of an object, then, is the product of its mass and its specific heat capacity. Thus, if m, C, T_f, and T_i refer to mass, specific heat capacity, the final temperature, and the initial temperature, the heat transferred (H) to an object is

$$H = mC(T_f - T_i)$$

When a pure material absorbs heat while undergoing a change of state, there is no change in temperature. The amount of heat required per gram of material is called the *heat of fusion* or *heat of vaporization*, as the case may be.

A calorimeter is a device in which quantities of heat can be produced (as from a chemical reaction or by passing electric current through a resistance) or simply transferred (from a hot body to a cold one) without significant exchange of heat with the outside. When the requirements are not severe, the exchange of heat may be limited by simple insulation. More sophisticated devices employ silvered vacuum jackets similar to the familiar Thermos bottle. For the ultimate in precision, the vacuum-jacketed calorimeter may be immersed in a controlled temperature bath that automatically follows the temperature inside the calorimeter, thus almost completely eliminating the flow of heat. A closed system that does not exchange heat with its surroundings is said to be *adiabatic*.

In the adiabatic transfer of heat, the heat lost by the hot body must be exactly equal to the heat gained by the cold one, and the final temperature of both bodies must be the same. Thus,

$$m_1 C_1 (T_1 - T_3) = m_2 C_2 (T_3 - T_2)$$

where the Subscript 1 refers to the hot body, Subscript 2 refers to the cold body, and T_3 is the final temperature.

PROCEDURE

Part 1: Heat of fusion of ice

A calorimeter sufficient for this experiment can be made by wrapping a paper towel around a 250-ml beaker and inserting this inside a 400-ml beaker. (A better calorimeter can be made from two styrofoam coffee cups, one inserted inside the other.) A square cardboard with a hole through the center to accommodate the thermometer serves as a cover. The special thermometer with divisions of tenths of a degree is an expensive and fragile instrument, and since it will serve also as a stirrer, it must be protected. Obtain a 1/4-inch and a 1/8-inch section of rubber tubing. After moistening the thermometer, position the 1/4-inch section on the thermometer above the cardboard so that, when the thermometer is inserted through the cover, the bulb cannot reach the bottom of the beaker. Then place the smaller section of tubing around the bulb itself to prevent the bulb from accidentally hitting the side of the beaker. As an added precaution, lay the thermometer on a paper towel to cushion it when laying it on the laboratory desk.

Practice holding the cover with one hand, supporting the thermometer between two fingers, and rotating the cover on the top of the beaker to obtain a gentle stirring action. Hold onto the beaker with the other hand.

The fundamental unit of heat, the calorie, is defined to make the specific heat capacity of water at approximately room temperature very close to unity. In this experiment, ice is added to a quantity of water in the calorimeter and the decrease in temperature is observed. Hence, the amount of heat supplied to the ice to melt it can be calculated. However, not only does the water supply heat, but so does the calorimeter itself. Consequently, the heat capacity of the calorimeter, C_{cal}, must be determined.

Heat capacity of the calorimeter

Adjust the temperature of some tap water to within a degree of 25°C, and place exactly 100 ml (m_1) of this in the calorimeter. Adjust a support for the thermometer and cover the calorimeter. Place about 100 ml of water in an Erlenmeyer flask and swirl in an ice bath for a couple of minutes. After the water is close to 0°C, select several chunks of ice just small enough to get through the mouth of the flask, and continue to swirl the water for another minute. Read and record the temperature of the water in the calorimeter. When reading the thermometer, hold it vertically and read it with your eye level with the mercury thread to avoid parallax errors. Keeping an eye on a watch or clock, start the gentle stirring of the calorimeter. Stop stirring every 30 seconds and read the thermometer to the nearest quarter division. Do this for two minutes. Note the time and, holding your finger across the mouth of the flask to prevent any ice from leaving, pour the ice water into the calorimeter. Commence stirring and reading the temperature at 30-second intervals for a period of three or four minutes. Record the temperature in tabular form. Measure the total volume of water in the calorimeter. The total volume less the original volume is the added volume of ice water, which will be the approximate weight of the added ice water (m_2).

Plot your data, temperature versus time. Although the equalization of the water temperature is almost instantaneous, it is the rather slow response of the thermometer that necessitates measurement for several minutes. To correct for heat flow from outside the calorimeter, the steady upward drift of the plot can be extrapolated back to the time of mixing to get a more accurate temperature change. Considering that the "hot body" is both the 100 g of water and the calorimeter, calculate the heat capacity of the calorimeter, C_{cal}

$$m_1 C(25 - T_3) + C_{cal}(25 - T_3) = m_2 C(T_3 - 0)$$

Heat of fusion

Place 160 ml of water that is close to 25°C in the calorimeter and stir until a constant temperature reading is obtained. Record the data. Place an open paper towel on a platform balance and select about 40 g of larger chunks of ice. Record the weight. The wet towel will be weighed later. Pick up the corners of the paper towel and gently shake the ice around to absorb as much water as possible. Immediately transfer the ice to the calorimeter without splashing. Stir and record temperatures as before. Continue to read the temperature for a couple of minutes after it has stopped dropping. Now weigh the wet towel to obtain by difference the weight of ice.

Treat the temperature data as you did before. Calculate the heat of fusion of ice. Observe that the heat transferred is the sum of that quantity of heat required to melt m_2 grams of ice and that required to heat m_2 grams of water from 0°C to the final temperature

$$m_1 C(25 - T_3) + C_{cal}(25 - T_3) = H_f m_2 + m_2 C(T_3 - 0)$$

Part 2: Heat capacity of a metal

Obtain a piece of unknown metal from the instructor. Weigh the unknown on a triple beam balance. Add 100 ml of water at 25°C to the calorimeter. Heat the unknown in boiling water. Record thermometer readings of the calorimeter for two minutes. With tongs, transfer the metal to the calorimeter. Swirl the calorimeter very gently and read the thermometer every 30 seconds for three minutes, taking care not to hit the metal with the thermometer, or vice versa. Treat your data as before. Calculate the approximate atomic weight of the metal (from the rule of Dulong and Petit), as well as the specific heat of the metal.

Put the thermometer in its case and return it and the tubing sections to the stockroom.

SUPPLEMENTARY READING

Dickerson, Gray, and Haight (Chapter 14, Sections 1 and 6)

Masterton and Slowinski (Chapter 4, Section 1)

Sienko and Plane (Chapter 7, Sections 4 and 5)

Brescia, *et al.* (Chapter 3, Section 8)

Brown (Chapter 9, Sections 3 and 5)

Pauling (Chapter 5, Section 8)

Hammond, Osteryoung, Crawford, and Gray (Chapter 8)

For a discussion of Dulong and Petit see:

Dickerson, Gray, and Haight (Chapter 1, Section 8)

Masterton and Slowinski (Chapter 2, Section 3)

Mahan (Chapter 1, Section 4; Chapter 3, Section 5)

Brescia, *et al.* (Chapter 4, Section 8)

Brown (Chapter 3, Section 4)

Pauling (Chapter 5, Section 8)

QUESTIONS

1. Define heat capacity. Give a set of units for heat capacity.
2. How is the calorie defined?
3. Why is there a heat of fusion? Why must heat be added to convert a solid to a liquid at the same temperature?
4. Why does adding heat to a body without a change in phase (solid, liquid, or gas) generally result in an increase in temperature?
5. How can the law of Dulong and Petit be rationalized? (Why is it so?)

PROBLEM

In this experiment, the temperature effects occurring when "hot bodies" are brought into contact with "cold bodies" were studied. From this experience and your intuition, explain why it cannot be possible to reach absolute zero temperature completely by such processes.

EXPERIMENT 7

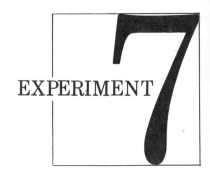

Equipment and supplies

One 250-ml Erlenmeyer flask

Pin or needle

One 600-ml beaker

Analytical balance

Barometer

Burner, wire gauze, ring stand

One piece of aluminum foil, 10 × 10 cm

Boiling stone

3 ml Unknown liquid

Time requirement

Two hours

MOLECULAR WEIGHTS BY VAPOR DENSITY

The purpose of this experiment is to find the molecular weight of a substance by direct measurement of its vapor density.

HISTORICAL NOTE

Democritus, Epicurus, Lucretius, and a few other ancient thinkers believed that matter was not infinitely divisible, but consisted of indivisible particles or atoms. However, during the Middle Ages and the Renaissance, Aristotle was considered the highest authority on nearly everything, and he had taught that matter was infinitely divisible. In the seventeenth century, the atomic theory was revived. Newton,[10] for example, thought that a gas consisted of small particles that repel each other. Expansion of the gas was explained by this repulsion.

Dalton,[11] however, was the first to make the atomic theory quantitative, and to apply it to chemistry. Dalton, like Newton, considered a gas to consist of particles or atoms separated by considerable space. The "atoms" or particles of compounds were composed of two or more elementary atoms, whereas elementary gases were believed to be single, simple atoms. Dalton concluded that all of the particles of any one gas are identical in size and weight, but different from the particles of any other gas. This conception implies a definite "atomic" weight for any substance and definite proportions for any compound substance; consequently, the laws of conservation of mass, of constant proportions, and of multiple proportions follow directly. The law of conservation of mass already had been proved experimentally by Lavoisier in 1774.

Dalton also constructed a table of atomic weights, all relative to an assumed value for hydrogen, $H = 1.0$. He assumed that in the binary compounds of lowest proportion, the atoms were combined in the ratio 1:1 ("law" of greatest simplicity); that is, that water was OH, and ammonia NH. With this assumption, and the composition of water by weight taken as eight parts of oxygen to one part of hydrogen (Dalton's analyses were faulty also; actually, he found the ratio of oxygen to hydrogen to be 5.66 to 1.0), an atomic weight of 8.0 was obtained for oxygen.

In 1808, Gay-Lussac[12] reported the results of gaseous reactions, among them that two volumes of hydrogen combined with one volume of oxygen to form two volumes of steam. These experiments were summed up as the law of Gay-Lussac: Gases always combine in volumes having simple ratios, and the volume of the product has a simple ratio to the volumes of the reacting gases.

Using Gay-Lussac's data, Berzelius observed (as Gay-Lussac had also) that the simplest explanation was that equal volumes of gases contain equal numbers of atoms. Therefore, water must be H_2O not HO, since two volumes of hydrogen combined with one of oxygen. This formula and the elemental analysis of water (hydrogen to oxygen 1:8, by weight) led him to assign atomic weights of H = 1.06 and O = 16.0 in his first table of atomic weights, published in 1813 (actually, he set O = 100; thus, H = 6.64). His table,[2] in addition to containing atomic weights based on the law of combining volumes, contained many that were based on analysis of the oxides of various elements. In these cases, he was forced to guess by analogy and intuition whether an oxide was MO, M_2O, MO_2, or so on.

The phlogiston theory only recently had been refuted. Consequently, chemists were very suspicious of theorizing, and especially of those theories that were not substantiated completely by experiment. Most chemists regarded Dalton's laws as merely brilliant speculation until Berzelius substantiated them with a large number of careful analyses. Similarly, they considered Gay-Lussac's law rash, substantiated by only a few experiments that they, including Dalton, believed to be no more exact than their own work (accuracy was not easy to achieve at this time). In fact, atomic weights also were considered very theoretical, most chemists preferring to use systems of combining weights that were simply the amounts of various substances reacting with amounts of various other substances. In addition, Gay-Lussac's work had some unexplained aspects. For example, why did two volumes of hydrogen and one volume of oxygen give exactly two volumes of steam, and why did carbon monoxide, which was known to be CO, have a smaller vapor density than oxygen, which was supposed to be just O.

In 1811, Avogadro[13] published an explanation of Gay-Lussac's work on the combining volumes of gases without any new experimental evidence. He pointed out that the simplest explanation for the simple proportions of reacting volumes was that equal volumes contained equal numbers of molecules, whether simple atoms or compounds. But he also pointed out that one volume of oxygen and two volumes of hydrogen producing two volumes of water vapor required that each oxygen molecule be divided into two parts upon combination with hydrogen. Therefore, the oxygen molecule must be composed of two (or multiples of two) particles. This also explains the relationship between the vapor densities of carbon monoxide and oxygen, which is the ratio of the molecular weights, 28:32. Avogadro's paper was ignored almost completely. In 1814, Ampère[14] published similar conclusions, derived mathematically.

One of the reasons why these ideas were not well received was the current theory of chemical affinity. After Davy's and Berzelius' work on electrolysis of salts, Berzelius developed the theory that the atoms in compounds were held together by electrical attraction of positive and negative charges. As a consequence of this theory, the union of two oxygen atoms that were both electronegative seemed impossible.

In 1819, Dulong and Petit[15] published their experiments on the specific heats of metals. They found that for solid elements, the product of the atomic weight and the specific heat was a constant, which they called the "atomic heat." This relationship became the law of Dulong and Petit, and could be used, together with analyses of compounds, to derive the correct atomic weights of elements.

In 1819, Eilhard Mitscherlich[16] reported his work on the isomorphism of crystals, which led to his law of isomorphism. He found that ferric alum and chrome alum, for example, have exactly the same crystalline shape, and concluded that chromium in one compound occupied the same place as iron in the other compound, the rest of the crystal remaining the same. Then, from analysis of the alums, the relative atomic weights of iron and chromium could be determined.

In 1819, Berzelius observed that if equal volumes of gases contain equal numbers of atoms, then the densities of the elementary gases must be proportional to their atomic weights. In 1826, Dumas[17] reported his method of measuring vapor densities of elements that are not gases at ordinary temperatures. In his method, of which this laboratory assignment is a simplified form, the substance is placed in a globe with a long, thin neck. The globe is heated at a certain temperature until the material is vaporized completely, then the neck is sealed. The globe is weighed and later weighed with water in it. Knowing the weight of vapor, and its pressure, volume, and temperature when it was vaporized, one can calculate the atomic weight relative to some standard, say hydrogen, $H = 1.0$. For iodine and bromine, Dumas obtained results in good agreement with those from chemical analyses. When Dumas measured the vapor density of mercury, he obtained an atomic weight of 99.45, which is half the value obtained by Berzelius ($Hg = 202.86$) from chemical data based on analogy and the law of Dulong and Petit. Later, he experimented with sulfur and obtained an atomic weight of 93.74, which contradicted the result derived from the law of Dulong and Petit, as well as the best analogy and Mitscherlich's law of isomorphism. Dumas finally concluded that sulfur vapor at 230°C had three times as many atoms in a molecule as oxygen has. Mitscherlich pointed out later that mercury must have half as many atoms in a molecule as oxygen. Berzelius became disillusioned with the vapor density method for obtaining atomic weights, because if the number of atoms in a molecule of an element could not be determined, then the atomic weight could not be obtained.

In 1834, Faraday[18] announced his law of electrochemical action (see Experiment 9): The chemical power of a current of electricity is directly proportional to the absolute quantity of electricity that passes. By measuring the amount of a metal dissolved or plated out, relative to one gram of hydrogen displaced, he arrived at a series of "atomic" weights. Actually, they were what are now called equivalent weights; consequently, in some cases, they were submultiples of those values in Berzelius' table of atomic weights.

In the first half of the nineteenth century, great advances were made in organic chemistry. But by 1860, formulas for water ranged from HO to H_2O to H_2O_2, and the large number of formulations for the more complex organic compounds was grotesquely comic. This confusion resulted from the discrepancy in the atomic weight of oxygen (8 or 16). To decide what to do, a congress was convened at Karlsruhe, in September 1860. At this meeting of more than 100 prominent chemists, Stanislao Cannizzaro, professor at Genoa, eloquently explained Avogadro's 50-year-old hypothesis. He also gave out reprints of an earlier paper of 1858[19] that apparently clinched the argument. After this, all chemists accepted a single atomic theory with a single set of atomic weights. This laid the groundwork for the discovery of the periodicity of the elements that followed soon after.

PROCEDURE

Weigh a 250-ml (or 125-ml) Erlenmeyer flask together with a small sheet of aluminum foil (about 6 × 6 cm) on an analytical balance. The flask must be clean and dry. Record the weight of the foil and flask filled with air.

Obtain a sample of an unknown liquid. Pour about 3 or 4 ml of the sample into the flask. Crimp the aluminum foil over the top of the flask as tightly as possible. If a short length of thin copper wire is available, it may be twisted around the neck of the flask over the aluminum foil to make the seal better. Puncture the foil with a pin to make a tiny hole.

Clamp the flask so that it is up to the neck in a large beaker half full of water. Add three or four boiling stones to the beaker. Do not get water under the aluminum foil! Heat the water, and boil it for about two minutes after the disappearance of all drops of liquid from the flask. Do not reduce the temperature of the flask by removing it from the bath or by letting the bath cool during the vaporization of the liquid, because air will reenter the flask. When the liquid has been gone two minutes, stop heating, and remove the flask from the beaker. Cool the flask under a stream of water from the tap. Take care not to get the aluminum foil wet. Dry the outside of the flask well with a towel, and weigh it on an analytical balance. Record the weight of the flask and foil plus the condensed liquid and air. The condensed liquid is of negligible volume, so the weight of air in the flask is practically the same as in the first weighing. Therefore, the difference between the weighings is the weight of liquid that filled the flask as a vapor at the boiling point of water.

Record the barometric pressure and the temperature to which the flask was heated. The boiling point of water at this pressure may be obtained from a handbook.

Now fill the flask with water. Weigh the flask (and aluminum foil) filled with water on a triple beam balance. Do not use the analytical balance because the weight will exceed the load limit of the balance; also, such precision is not needed. Record the weight and the temperature of the water. Find the density of water at this temperature from a handbook, and calculate the volume of the flask.

From the measured pressure, volume, temperature, and weight of the unknown liquid when it filled the flask, assuming that the vapor is an ideal gas, calculate the molecular weight of the liquid. The value will be approximate not only because of the limitations of the apparatus, but also because the vapor is just above the boiling point of the liquid, a range where nonideal behavior often is observed.

SUPPLEMENTARY READING

Dickerson, Gray, and Haight (Chapter 1, Section 7)

Hammond, Osteryoung, Crawford, and Gray (Chapter 1, Section 8)

QUESTIONS

1. If the liquid did not completely vaporize in the experiment, would the apparent molecular weight be larger, or smaller, than it should be?
2. In the weighing after the vaporization of the liquid, what effect would water on the outside of the flask or under the aluminum foil make in the weight of condensed vapor and in the calculated molecular weight?
3. Would the error in Questions 1 and 2 be larger, or the same, as the errors that might be made in measuring the temperature of the flask, the barometric pressure, or the volume of the flask?
4. What principle or law of nature permits us to calculate the molecular weight of a gas from the weight of a certain volume of its vapor?

PROBLEMS

1. The rare gases helium, neon, argon, krypton, and xenon generally do not form stable compounds. Xenon, however, has recently been found to form oxides and fluorides, for example. In 1900, Ramsay and Travers reported the discovery of argon and the measurement of the density of the gas. If the gas density is 1.783 g liter^{-1} at standard conditions, show how the gram atomic weight can be calculated.
2. A compound that has the composition BH_3 is found to have a vapor density of 1.24 g liter^{-1} at 0°C and 760 torr. Calculate the molecular weight. Also write the molecular formula.

EXPERIMENT 8

Equipment and supplies

One apparatus (or commercial equivalent from Welch Scientific Co., for instance)

One 400-ml beaker

Burner, wire gauze, ring stand

Thermometer

Barometer

Ice

Dry Ice

Time requirement

Two hours

CONSTANT VOLUME GAS THERMOMETER

This experiment is an investigation of the relationship between pressure and centigrade temperature for a gas that is confined to a constant volume. The relationship thus found will be used to measure the sublimation temperature of solid carbon dioxide. The data also will be used to devise the absolute temperature scale.

HISTORICAL NOTE

The first edition of Robert Boyle's book *New Experiments, Physico-Mechanical, Touching the Spring of the Air and Its Effects* was published in 1660. In the second edition in 1662, he described experiments that now are summed up as Boyle's law: The volume of a fixed amount of a gas varies inversely as the pressure, if the temperature remains constant. In 1802, J. A. C. Charles communicated to Gay-Lussac the results of experiments done in 1787, showing that the volume of a fixed amount of a gas varies directly with temperature, if the pressure is constant.[20] Also in 1802, Gay-Lussac[20] reported that the thermal expansion of various gases could be described by the equation (Gay-Lussac's law)

$$V_\Theta = V_0(1 + \Theta\alpha)$$

where V is the volume at $\Theta°C$ and $0°C$, and α is the fraction of volume change per degree Celsius. Gay-Lussac found α equal to 1/267. In 1847, the French chemist and physicist Henri Regnault refined the value of α to 1/273, using improved procedures. William Thomson (later Lord Kelvin) recognized that an absolute zero of temperature was implied in Charles' law and in Gay-Lussac's law, and defined the absolute temperature scale in 1848.[21] Temperature, T, on the new absolute scale is

$$T = \Theta + \frac{1}{\alpha}, \text{ or } T = \Theta + T_0$$

where Θ is temperature of the Celsius scale and T_0 (or $1/\alpha$) is 273. Combining this with Gay-Lussac's law

$$V = V_0(1 + \Theta\alpha) = V_0\left(1 + \frac{\Theta}{T_0}\right) = V_0\left(1 + \frac{T - T_0}{T_0}\right)$$

$$= V_0 + \frac{V_0 T}{T_0} - \frac{V_0 T_0}{T_0} = \frac{V_0 T}{T_0}$$

results in the now familiar way of writing Charles' law

$$\frac{V}{V_0} = \frac{T}{T_0}$$

As early as 1822,[22] Gay-Lussac recommended air thermometers for precise work, because mercury-in-glass thermometers were less dependable. Heating and cooling caused the glass of that time to expand and contract, changing shape irreversibly, thus leading to nonreproducible temperature readings. The practical constant-volume gas thermometer was developed about 1847 by Regnault,[23] and was used for precise measurements in the nineteenth century.

PROCEDURE

Part 1: Measurement of atmospheric pressure

Examine the barometer. A simplified diagram of a barometer is shown in Figure 8-1. The barometer consists of a glass tube filled with mercury, a reservoir at the bottom, and a scale for measuring the height of the mercury column. The space above the mercury column in the glass tube is a vacuum, so there is no force pushing down on the top of the column of mercury. The column of mercury is supported by the pressure of the atmosphere on the surface of the mercury in the reservoir. Notice the ivory pointer in the reservoir of the laboratory barometer, and the thumb screw by which the mercury level can be adjusted to touch the pointer. This ivory pointer is the lower end of the millimeter scale that measures the height of mercury. Only the portion of the scale between about 600 and 800 mm is ruled. If you do not understand how to read the vernier scale at the top, you may consult the instructor. However, it is not necessary to read the barometer closer than about 1 mm for this experiment, and this can be done with the eyes only.

Adjust the mercury in the reservoir level with the tip of the pointer. Keeping your eyes level with the lower part so that the front and back edges coincide, adjust the movable meniscus. Read directly across to obtain the height of the mercury column that is being pushed up by atmospheric pressure. Record this reading.

Part 2: Calibration of the gas thermometer

Examine the gas thermometer apparatus. Figure 8-2 is a drawing of such an apparatus. Since one side of the manometer is open to the atmosphere, the difference in heights of the two mercury columns rep-

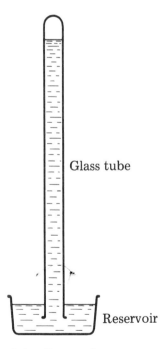

Figure 8-1. Barometer (simplified).

Procedure 71

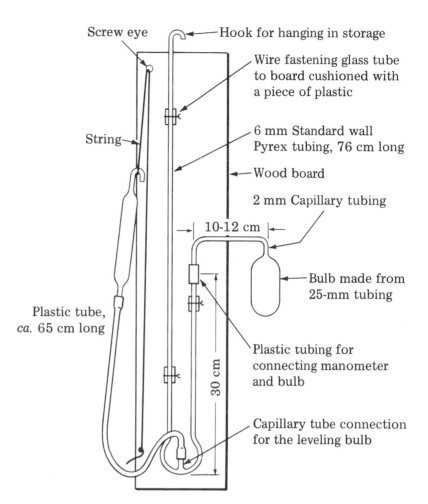

Figure 8-2. The apparatus for the experiment is constructed of 6-mm hard glass tubing connected to a bulb by a capillary tube. To connect the bulb, heat the glass tubing (it is best to use a heat gun), then push the plastic tubing on while the glass is hot. Similar apparatus is available commercially.

resents a difference in pressure between atmospheric pressure, which you have just measured, and the pressure in the closed side. Lift the leveling bulb slightly and observe the behavior of both columns of mercury. *Care must be exercised throughout this experiment to avoid letting the mercury level in the closed side get either so high that the mercury might spill over into the gas tube or so low that the gas might escape through the bottom.* This can happen only through inattention. A change in the volume of the gas will make it necessary to repeat the entire experiment and possibly to reassemble the apparatus. As the temperature of the gas is changed it will be necessary to raise and lower the bulb by pulling the supporting string so that the closed-side mercury level is kept close to the middle of the capillary tube.

There is a mark at the top of the manometer tube on the closed side. Whenever the mercury level is brought exactly to the mark, the enclosed volume of gas is the same. The mark should be close to the glass capillary tube connected to the bulb so that the volume of gas outside the bulb is as small as possible. However, do not make the mark at the glass capillary, because then it will be difficult to avoid getting mercury into the bulb.

Arrange a ring stand, ring, and wire gauze to support a 400-ml beaker as high up around the bulb of the gas thermometer as possible. Fill the beaker two-thirds full of ice, add distilled water, and stir. Position the beaker around the tube. Suspend a mercury thermometer in the ice-water mixture. Add more water to fill the beaker and submerge the tube. Stir the ice until the temperature is constant at, or very near, 0°C.

Rule three columns in your notebook and label them temperature, manometer reading, and total pressure. *Slowly* raise the leveling bulb until the mercury just reaches the mark on the closed side. Read and record the difference in height of the mercury in the open side. Remember to tap the manometer. Now stir the ice for about a minute and read the pressure again. If the second measurement does not agree with the first, stir and repeat. Read the mercury thermometer. If the thermometer reading is not 0.0°C, the thermometer is in error. Record the actual reading.

Empty the ice and refill with distilled water, adjusted to about 25°C. After waiting about two minutes for temperature equilibration, measure the temperature and pressure. Check as before.

Adjust the Bunsen burner to a small, soft flame and heat the water to about 50°C. Avoid burning the manometer board! Also, try not to heat the plastic tubing joint. Heat will soften it, and may cause it to leak, thus spoiling the experiment. When the correct temperature has been reached, stir the water and heat intermittently to maintain a constant temperature. Record the temperature and pressure. Make a similar measurement at about 75°C.

Bring the water to a boil. After allowing for equilibration, measure the pressure again after a minute. Remove the burner and allow the water to cool. *Remember to check the mercury level now and then while the gas is cooling and adjust the bulb as necessary.*

The true boiling point of water near 100°C is given, in degrees Celsius, by the equation

$$t = 100.00 + 0.037(p - 760)$$

at the atmospheric pressure p. Calculate this temperature. If it does not agree with your thermometer reading, it is the thermometer that is in error. If the discrepancy at the ice point or the boiling point is larger than 0.2°C, ask the instructor to help you correct your temperature readings. Smaller discrepancies may be ignored.

Part 3: Measurement of the pressure of the gas at the sublimation temperature of carbon dioxide

When the water in the 400-ml beaker is cool enough, remove that beaker and replace it with a dry 150- or 250-ml beaker. Remove the mercury thermometer, since the low temperature encountered in the following operation would freeze the mercury and might damage the thermometer. Adjust the leveling bulb so that the mercury level is close to the bottom on the closed side. Using a paper towel, obtain a scoop of crushed, solid carbon dioxide, commonly called Dry Ice. Use care in handling the Dry Ice as it can cause a painful burn. Shake a small amount of the Dry Ice into the beaker to cover the gas tube. After allowing for temperature equilibration, measure the pressure. Check the pressure measurement a minute later. When the measurement is complete, remove the beaker. Watch the mercury level until the tube is back to room temperature. Adjust the mercury level to the middle of the closed-side capillary. Return the Dry Ice. Record the pressure.

CALCULATION AND REPORT

Make a graph of the pressure in the bulb (on the vertical axis) versus temperature in degrees Celsius (on the horizontal axis). The temperature scale should extend from $-300°C$ to $+100°C$, and the pressure should begin at zero. Draw a line through your five measurements. If the points do not lie on a straight line, or nearly so, check your calculations.

Write an equation for this line (similar in form to the law of Gay-Lussac), and include it in your report.

To determine the sublimation temperature of carbon dioxide, extend the line on your graph to the observed pressure reading at the Dry Ice temperature. Read the corresponding temperature from the graph. This is the temperature of sublimation found for carbon dioxide.

You just have used the experimental pressure-temperature relationship to measure a temperature well below the range of a mercury thermometer (mercury freezes at $-39°C$). Extrapolate your linear plot of pressures to zero pressure. To what temperature does zero pressure correspond? This value is very near absolute zero.

In fact, nitrogen condenses to a liquid near $-200°C$ (boiling point at one atmosphere is $-196°C$), so the pressure would drop drastically because a liquid occupies much less volume than a gas. Other gases liquefy at other temperatures. Extending your line smoothly to zero pressure corresponds to the behavior of a gas that does not liquefy. Nevertheless, it is found that the pressure-temperature relationship is the same for all gases, provided that the pressure is low. All *P-T* data can be extrapolated to the same absolute zero.

Convert each temperature used in calibrating the gas temperature ($0°C$ to $100°C$) to an absolute scale, using your experimental absolute zero. Prepare a table listing the Celsius temperature, your absolute temperatures, the observed pressures, and the ratio of the observed pressure divided by the absolute temperature (P/T). Express the result, which is apparent from the ratios, as an algebraic equation.

SUPPLEMENTARY READING

Dickerson, Gray, and Haight (Chapter 2, Sections 4 and 5)

Masterton and Slowinski (Chapter 5, Section 4)

Mahan (Chapter 2, Section 1)

Brescia, *et al.* (Chapter 2, Section 3)

Brown (Chapter 3, Section 2)

Hammond, Osteryoung, Crawford, and Gray (Chapter 2, Sections 1-3)

QUESTIONS

1. How does the pressure change when the temperature is increased, keeping the volume and amount of gas constant? Rationalize this phenomenon in terms of the kinetic theory of gases (random motion of point masses).
2. What physical significance does absolute zero temperature have for an ideal gas?
3. Does it matter in this experiment what the gas is in the bulb?
4. Suggest a reason why a gas thermometer might be more sensitive for very small temperature changes than a mercury thermometer.

PROBLEM

From your experimental observation of the change in pressure with change in the temperature of a gas, calculate what temperature a steel cylinder of nitrogen can withstand without exceeding the tested safety limit of 6000 lb in.$^{-2}$. At 25° C, the cylinder contains nitrogen at 2250 lb in.$^{-2}$ pressure.

EXPERIMENT 9

Equipment and supplies

One copper anode (about 3.5 × 1 × 1/16 inch, cut from bar stock in 8-foot lengths)

One ammeter (Parker Model S 35 is especially good since it has a fuse)

One 6 V storage battery

One alligator clip for connection to the anode

Three pieces of insulated copper wire (one stripped bare at the end for the cathode)

One 50-ml burette

One 3-inch piece of rubber tubing

One clamp and ring stand

One pipette bulb

Clock with second hand

Barometer

Thermometer

Analytical balance

25 ml $3M$ Sulfuric acid

Time requirement

One and one-half hour

ELECTROLYSIS OF COPPER; THE FARADAY

The purpose of this experiment is twofold: first, to verify the stoichiometry of an electrolysis reaction; and second, to measure the faraday and then determine Avogadro's number.

HISTORICAL NOTE

In 1800, Alessandro Volta[24] reported the generation of electricity from a pile of alternating plates of silver and zinc, separated by cloth soaked in brine. Later in the same year, Nicholson and Carlyle reported the decomposition of water by electric current into hydrogen and oxygen. A few years later, Davy isolated the elements sodium and potassium by electrolysis of their fused salts. Solutions of salts also were decomposed with electric currents.

In the *Philosophical Transactions of the Royal Society* (London) of 1831-1834, appear reports by Michael Faraday of his experiments in electricity and electrochemistry.[25] Faraday reported the first method of quantitatively measuring amounts of electric current. He invented a device that he called a "volta-electrometer," in which the volume of hydrogen evolved by passing the current could be measured. He reported that water is electrolyzed to hydrogen and oxygen when platinum electrodes are used, but when copper electrodes (or tin or iron) are substituted, no oxygen is formed. Only hydrogen appears at one electrode, and the other copper electrode dissolves. By measuring the amount of a metal dissolved or plated out for a standard volume of hydrogen formed, it was possible to obtain a series of equivalent weights, all based on the same criterion. The results of these experiments were condensed to what are now called Faraday's laws of electrolysis: The chemical power of a current of electricity is in direct proportion to the absolute quantity of electricity that passes.

Note: Current is measured in amperes, charge is measured in coulombs. An ampere is that current that carries one coulomb of charge past any point in each second. Thus, charge is equal to the product of current and time (in seconds). A current of 4 amperes flowing for 2 minutes delivers a charge of 480 coulombs. The faraday is the charge of one mole of electrons, that is, 96,484 coulombs. The charge on one electron is 1.602×10^{-19} coulomb.

78 Electrolysis of copper; the faraday

PROCEDURE

Obtain a copper anode (about 3 1/2 × 1 × 1/16 in.) and weigh it to the nearest milligram. Fill a 250-ml beaker half full with distilled water, and add 25 ml of $3M$ H_2SO_4.

Place the beaker of acid under the gas burette and immerse the burette so that the lowermost mark is even with the liquid level. *Carefully* suck the acid up into the burette, using a rubber pipette bulb. DO NOT use your mouth. Close the rubber tube at the top of the burette tightly with a clamp. If the burette is filled to the very top, it will be necessary to measure the volume of the burette above the last mark, but this can be done later. If the space above the last mark is left filled with air, it is necessary to correct for the compression of this volume of gas during the course of the experiment.

Study the diagram of the electrolysis setup (Figure 9-1), paying special attention to the anode, cathode, and signs of the battery terminals. Electrons leave the battery from the negative terminal. Position the cathode so that all of the bare part of the wire is well inside the burette. Attach the anode to the clip, but do not immerse it in the acid yet. Place a paper towel beside the beaker. Check your wiring. The ammeter fuse will blow if you make a mistake.

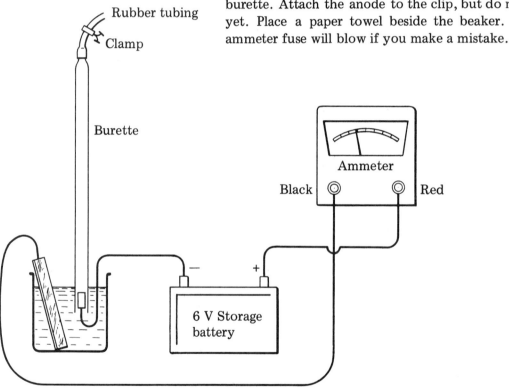

Figure 9-1. Apparatus for Experiment 9 using Parker ammeter Model S 35 with fuse (1-1/2% accuracy).

Decide on a specific time to start, and record this time. Use a watch with a second hand, or the wall clock. At the start time, immerse the anode in the solution. If the current is too large (off the ammeter scale), dilute the acid solution in the beaker. Do not touch or jar the anode as this will change the current. Read the current on the ammeter after *15 seconds* have elapsed, and continue to read the current at exactly 30-second intervals, listing your readings in a neat column. Stop the electrolysis by removing the anode just before the water level in the burette gets to the lowermost mark. Do not overshoot. As you remove the anode, read the time. Lay the anode on the towel. Record the termination time.

Record the burette reading, taking care that the water levels inside and outside the burette are the same. Remove the beaker of acid to your desk where you can measure the temperature. Clean around the apparatus for the next person.

Read the barometric pressure and subtract 3 mm from the reading to correct the barometer to its standard condition at zero degrees. Read the temperature of the acid solution. Assume that this is also the gas temperature. Obtain the vapor pressure of water at this temperature from a handbook or other reference book. Compute the partial pressure of the hydrogen gas. Compute the volume of dry hydrogen at standard temperature and pressure, and the number of moles of hydrogen.

Weigh the anode after you have washed and thoroughly dried it. Compute the number of moles of copper consumed.

Compute the ratio of moles of hydrogen produced to moles of copper consumed. Attribute any deviation from the theoretical value to error, and state your accuracy as a percentage of deviation.

Calculate the average of all of your current readings. Calculate the time in seconds. Calculate the charge transferred. From the number of moles of copper dissolved, compute the number of moles of charge transferred. Compute the charge of one mole of electrons, that is, the faraday. Compare this value with the accepted value and state your percentage of error.

Also compute Avogadro's number, knowing the charge of one electron.

SUPPLEMENTARY READING

Dickerson, Gray, and Haight (Chapter 3, Sections 1-5)

Masterton and Slowinski (Chapter 20, Section 3)

Sienko and Plane (Chapter 13, Sections 1-3)

Mahan (Chapter 7, Section 7)

Brescia, *et al.* (Chapter 15, Sections 1-4, 11 and 13)

Brown (Chapter 11, Sections 4 and 5)

Pauling (Chapter 8, Sections 5-7)

Hammond, Osteryoung, Crawford, and Gray (Chapter 4, Section 2)

QUESTIONS

1. Describe the reactions taking place at the electrodes in this experiment. Write balanced equations for the reactions.
2. Describe processes that occur when electricity is conducted through the acid solution (in addition to the electrode reactions, noting that these reactions either consume or produce ions at the electrodes).

3. If the copper electrode was wet when it was weighed after electrolysis, how would the value determined for the faraday be affected?

PROBLEM

Two cells having two platinum electrodes each are connected in series to a source of direct electric current. One cell contains aqueous silver nitrate and the other contains aqueous copper sulfate. Calculate the ratio of silver plated out to copper plated out.

EXPERIMENT 10

Equipment and supplies

One 1-ml pipette or 5-ml Mohr graduated pipette

Pipette bulb

One dish (Pyrex pie plate, large watch glass or Petri dish)

One glass rod (longer than width of dish)

One sheet of glass, 10 × 10 inches

One grease pencil

Four 150-mm test tubes

One 10-ml graduated cylinder

Analytical balance may be used

1 ml Oleic acid

40 ml Pentane

Grease

0.1 ml Piston oil (motor oil heated 8 hours at 300°C)

0.1 ml Dilute (6M) hydrochloric acid

Lycopodium powder or cork dust (optional)

Time requirement

One hour

DETERMINATION OF AVOGADRO'S NUMBER

The purpose of this experiment is to measure Avogadro's number directly.

HISTORICAL NOTE

In 1811, Avogadro[13] published the hypothesis that, in the gaseous state, equal numbers of molecules occupy equal volumes at the same pressure and temperature. However, it was not possible at that time to calculate the actual number of molecules in a cubic centimeter of a gas. Later, the number of molecules in a mole was calculated. Because Loschmidt was the individual who first calculated this quantity, Avogadro's number, the number of molecules per mole, is sometimes called "Loschmidt's number." Loschmidt's calculation was based on the kinetic theory of gases that had been developed by Maxwell and Clausius, among others. The mean free path, which is the average distance traveled by a moleclule before it collides with another molecule, is the average velocity of the molecules divided by the collision frequency of the molecules. This yields the following relationship:

$$\lambda = \frac{1}{\sqrt{2}\pi L S^2}$$

From this equation, L, Loschmidt's number per cubic centimeter, can be calculated if the mean free path (λ) and the diameter of a molecule (S) can be determined. The diameter of a molecule (S) can be obtained by assuming that the volume of the gas molecules themselves (not the empty space that the gas occupies) is equal to the volume of the condensed liquid, where they are in contact. The mean free path can be obtained experimentally from measurement of the viscosity of the gas (η) by the following relationship derived by Maxwell:

$$\eta = \frac{1}{3}\rho \bar{c} \lambda$$

where ρ is the gas density, and $\bar{c}$ the average velocity of gas molecules. The best value of Avogadro's number obtained by this method was 4.4×10^{23} molecules per mole.

Avogadro's number has been calculated by many other methods with varied results.[26] The relationship of the faraday ($\mathcal{F}$) to the charge of the electron (e) provides an important way of calculating the number of particles (N) in the mole of particles of electricity

$$\frac{\mathcal{F}}{e} = N$$

The faraday had been measured accurately in the nineteenth century. Consequently, in 1909, after Millikan determined the charge of the electron by the famous oil drop experiment, it was possible to calculate the value of 6.06×10^{23} molecules per mole for Avogadro's number (1917).

In 1941, R. T. Birge (a physicist at the University of California, Berkeley) calculated the present value of Avogadro's number from crystallographic data as $(6.0228 \pm 0.0011) \times 10^{23}$ molecules per mole, which was revised in 1945[27] to $(6.02338 \pm 0.00043) \times 10^{23}$. In making the calculation on sodium chloride, for example, the following equation was used:

$$N = \frac{Mf}{\rho \phi a^3}$$

where M is the molecular weight obtained from the best values of the atomic weights; f, the number of molecules in the sodium chloride unit cell ($f = 4$ for face-centered cubic); ρ, the density of the sodium chloride; and a, the length of the side of the unit cell. The factor ϕ is a geometric factor that is unity for a cubic unit cell. The length of the side of the unit cell (a) is obtained, using Bragg's law, from x-ray diffraction experiments. The wavelength of the x rays used could be determined accurately and directly by the use of ruled gratings, which were a recent development in 1941.

Although the method of determining Avogadro's number used in this experiment has not been used as a serious method for the measurement of N, a reasonable result is obtained. The experiment is based on an article by L. C. King and E. K. Neilsen.[28]

INTRODUCTION

In this experiment a very small quantity of oleic acid is permitted to spread over the surface of water as a film. Knowing the volume of oleic acid present, and the maximum area it covers on the water, we can calculate the thickness of the acid film. We can assume that this film is one molecule thick. With the aid of one further assumption, namely, that the shape of the molecule is known, we can calculate the number of molecules present in the film. Then, from the known molecular weight of oleic acid and its density, we can calculate the number of molecules present in a mole of the acid.

To measure a very small quantity of oleic acid, the oleic acid is dissolved in pentane. Pentane is a volatile (and also flammable!) solvent. When the dilute oleic acid-pentane solution is placed on the surface of the water, the pentane evaporates (almost completely), leaving the minute quantity of oleic acid.

PROCEDURE

Place 10 ml of pentane in each of four test tubes. To the first tube, add 1 ml of oleic acid and mix the contents by gently swirling the tube. Transfer 1 ml of solution from Tube 1 to Tube 2 and mix as before. Transfer 1 ml of the solution from Tube 2 to Tube 3, and mix. Finally, transfer 1 ml of solution from Tube 3 to Tube 4, and mix. The volume of oleic acid now present per milliliter of solution in Tube 4 can be calculated

$$\frac{1 \text{ ml}}{11 \text{ ml}} \times \frac{1}{11} \times \frac{1}{11} \times \frac{1}{11} = \frac{7 \times 10^{-5} \text{ ml oleic acid}}{\text{ml solution}}$$

A Pyrex pie plate makes a suitable dish for this experiment. The essential consideration is that the dish be large, about 8 inches in diameter, and clear and colorless because the film is not easy to see. Large watch glasses or Petri dishes can be used. The edge of the dish should be greased so that water does not wet the glass.

Fill the dish with water until the water level is just *above* the edge of the dish *everywhere* around the edge of the dish (Figure 10-1). The surface of the water then appears flat, but it is imperceptibly convex. Consequently, the film will tend to spread out. If the water is not above the edge of the dish everywhere, the surface of the water will be imperceptibly concave, therefore the film will be confined to the low spot in the middle. The dish can be leveled by putting paper under the low side if necessary. Add a drop of dilute hydrochloric acid to the water in the dish to prevent reaction of the oleic acid with any basic impurity in the water. Now clean the surface of the water by sweeping with a glass rod held horizontally.

Figure 10-1. Dish filled with water. The level of the water is exaggerated.

Add one *very small* drop of piston oil to the surface of the water with a fine medicine dropper or a small glass rod. (Piston oil is prepared by heating a good grade of motor oil to 300°C for 8 hours, or until it spreads as a monolayer.[31]) Let the piston oil spread out in a film. This film is slightly colored. Although it is unnecessary, some experimenters find that dusting the piston oil film with lycopodium powder or cork dust improves the visibility of the oleic acid film.

Now place exactly 0.10 ml of the most dilute (Tube 4) pentane solution of oleic acid on the surface of the water. Use a small graduated pipette or a calibrated medicine dropper for this transfer. The pentane evaporates and the oleic acid spreads out in a film that is distinctly visible, compared to the darker piston oil film around it, when the dish is viewed on a clean black bench. If the film of oleic acid reaches the edge of the dish, it will be necessary to clean the surface with a glass rod, add a drop of piston oil, and repeat the experiment with a smaller volume of pentane solution.

Support a sheet of glass (about 9-10 inches square) above the water on corks or on other convenient objects. With a grease marking pencil, quickly outline the shape of the oleic acid film as accurately as possible. Trace the outline of the film onto a clean sheet of paper. Then, with scissors cut out the film shape and weigh it on an analytical balance. Cut out a 10.0 × 10.0 cm square of the same clean paper and weigh it accurately also. Calculate the area of the film from these two weights. An alternative method of obtaining the film area is to trace the outline on graph paper and count the squares.

CALCULATIONS

The density of the oleic acid is 0.895 g ml^{-1}. Find the volume of oleic acid in the film. From the volume and the area of the film, calculate the thickness of the film. Assuming that the oleic acid molecule is a cube and that the film thickness is the length of one side of this cubic molecule, calculate the volume of one molecule. The gram molecular weight of oleic acid ($C_{18}H_{34}O_2$) is 282.5 g mole^{-1}. From the density and the weight of one mole, calculate the number of molecules in a mole, namely Avogadro's number.

SUPPLEMENTARY READING

Dickerson, Gray, and Haight (Chapter 1, Section 6)

Hammond, Osteryoung, Crawford, and Gray (Chapter 2, Section 4)

QUESTIONS

1. In this experiment, if it is assumed that:
   a) molecules are discrete and identical particles of matter,
   b) molecules are at contact distance in the liquid film, and
   c) the thickness of the thinnest film is of the order of magnitude of the molecular dimensions,

the number of molecules in the film can be estimated, if the volume of the film is known. If the weight of the film and the molecular weight of the liquid (defined relative to some standard) are known, then the number of molecules in a gram molecular weight (a mole) can be determined. What reasons can you give for the first two assumptions from your knowledge of chemistry?

PROBLEMS

1. Discuss sources of determinate experimental errors that might be important in this experiment. Discuss the consequences of the film being more than one layer of molecules thick.

2. The oleic acid molecule is not a cube. The molecule is a long chain of 18 carbon atoms with the oxygen atoms at one end

$$H_3C-CH_2-CH_2-CH_2-CH_2-CH_2-CH_2-CH_2-CH=CH-CH_2-CH_2-CH_2-CH_2-CH_2-CH_2-CH_2-COOH$$

The oxygenated end has the polar OH group that is better solvated by water (a polar solvent) than is the nonpolar carbon chain. Consequently, the oxygen end of the molecule points toward the water surface. Assuming that the carbon-carbon chain is arranged in a helix, the overall shape of the molecule would be a cylinder. Discuss what effect the error in assuming the cubic shape of the molecule makes in your answer. Without trying to calculate a new value for N, state how much difference (order of magnitude) the shape of the molecule will make.

EXPERIMENT 11

Equipment and supplies

One 25- × 200-mm test tube

One two-hole rubber stopper, No. 6

12-15 Inches glass tubing

3 Inches rubber tubing

24 Inches rubber or plastic tubing

One 500-ml flask (the filter flask can be used if the side tube is plugged with a cork)

One water trough

One funnel

One pinch clamp

Barometer

Thermometer

Clamp and ring stand

Analytical balance

Piece of unknown metal

30 ml Concentrated hydrochloric acid

Glycerol

Time requirement

One and one-half hour

2. The oleic acid molecule is not a cube. The molecule is a long chain of 18 carbon atoms with the oxygen atoms at one end

$$H_3C-CH_2-CH_2-CH_2-CH_2-CH_2-CH_2-CH_2-CH=CH-CH_2-CH_2-CH_2-CH_2-CH_2-CH_2-CH_2-COOH$$

The oxygenated end has the polar OH group that is better solvated by water (a polar solvent) than is the nonpolar carbon chain. Consequently, the oxygen end of the molecule points toward the water surface. Assuming that the carbon-carbon chain is arranged in a helix, the overall shape of the molecule would be a cylinder. Discuss what effect the error in assuming the cubic shape of the molecule makes in your answer. Without trying to calculate a new value for N, state how much difference (order of magnitude) the shape of the molecule will make.

EXPERIMENT 11

Equipment and supplies

One 25- × 200-mm test tube

One two-hole rubber stopper, No. 6

12-15 Inches glass tubing

3 Inches rubber tubing

24 Inches rubber or plastic tubing

One 500-ml flask (the filter flask can be used if the side tube is plugged with a cork)

One water trough

One funnel

One pinch clamp

Barometer

Thermometer

Clamp and ring stand

Analytical balance

Piece of unknown metal

30 ml Concentrated hydrochloric acid

Glycerol

Time requirement

One and one-half hour

THE COMBINING WEIGHT OF A METAL BY DISPLACEMENT OF HYDROGEN

The purpose of this experiment is to determine the combining weight, or equivalent weight, of an unknown metal by displacement of a measured volume of hydrogen. The equivalent weight of an element is the weight that will combine with, or displace, 1 g of hydrogen or 8 g of oxygen.

HISTORICAL NOTE

In 1766, Henry Cavendish[29] reported the first preparation of "inflammable air," although it probably was prepared by others earlier. He found that 254 grains of zinc in dilute sulfuric or hydrochloric acid produced 10 3/4 grains of inflammable air. He found, too, that this new gas was 10 1/2 times lighter than common air. He also prepared the gas by action of these acids on tin and iron.

In 1783, Lavoisier proved that this inflammable air was an element, and that it combined with oxygen to form water. For this reason, he named it *hydrogen*, water-former.

PROCEDURE

Set up a hydrogen generator (see Figure 11-1) consisting of a large (200- or 150-mm) test tube fitted with a two-hole rubber stopper. Into one hole fit a piece of glass tubing that reaches nearly to the bottom of the test tube. To prevent injury to yourself, read Sections 0.2 and 0.7; in particular, read the sections on glass cutting, fire polishing, and inserting tubes into rubber stoppers. Be sure to wrap a towel around the glass and to lubricate the stopper hole! To this glass tubing, attach a glass funnel, using a short length of rubber tubing. Put a screw or pinch clamp on the rubber tubing. Fit a delivery tube into the other hole in the rubber stopper. The hydrogen evolved will be collected by displacement of water from a 500-ml flask. Notice that the inner end of the delivery tube should be flush with the rubber stopper. Be sure that the apparatus is airtight. The water in the trough and cylinder must be within one degree of room temperature.

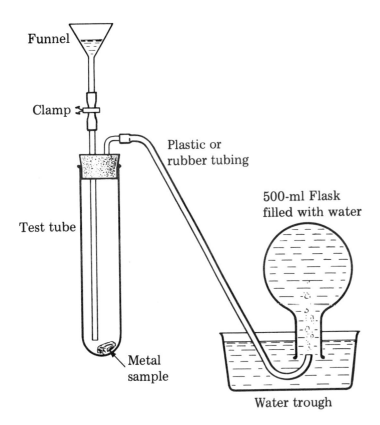

Figure 11-1. Hydrogen generator.

Ask for a sample of metal. Weigh it accurately. Put the sample in the test tube. Fill the whole system with water through the funnel, making sure that no air bubbles are present, and that the funnel is at least half full of water. See that the flask is in position to catch the hydrogen. Ask the instructor to inspect the apparatus. Include a simple diagram of the apparatus in your notes. Let the water level in the funnel go down to the stem of the funnel, and add about 30 ml of concentrated hydrochloric acid to the funnel.

Now allow the concentrated hydrochloric acid to flow slowly onto the unknown through the funnel. Be careful that no air enters the test tube. Clamp the rubber tubing tightly when the liquid level gets to the bottom of the funnel. If the liquid level in the funnel gets below the screw clamp, the whole experiment will be ruined. Allow the reaction to proceed until the metal sample is completely dissolved. The general reaction may be written

$$M + nH^+ \rightarrow \frac{n}{2}H_2 \uparrow + M^{n+}$$

Now run water through the funnel until all the hydrogen is displaced into the flask. If any air enters, it will be necessary to repeat the experiment. Hold the inverted flask in the trough so that the water level inside the flask is the same as that outside. If the water trough used for collection of the hydrogen is not large enough, use a larger trough or pail. See Figure 11-2, in which the flask is shown with the water levels equalized. When the water levels are equal inside and outside, the pressure of the atmosphere outside must equal the pressure of hydrogen and water vapor inside the flask. Then place your hand firmly over the mouth of the inverted flask to prevent escape of any water from the flask. Remove the flask from the trough, place it upright on the desk, and find the volume of the hydrogen by adding water from your graduated cylinder until the flask is full. Record the volume of hydrogen observed, and the temperature of the water, which must be within one degree of room temperature. Record the barometric pressure and the vapor pressure of water vapor at room temperature.

Calculate the volume of dry hydrogen at STP. Compute and record the equivalent weight of the unknown, recalling that 22.4 liters of hydrogen at standard temperature and pressure contain 1 mole of H_2 and weigh 2.016 g. Also recall that the equivalent weight of an element is the weight that will combine with, or displace, 1 g of hydrogen or 8 g of oxygen.

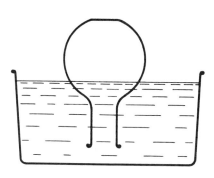

Figure 11-2. Equalizing the pressure.

SUPPLEMENTARY READING

Dickerson, Gray, and Haight (Chapter 1, Sections 4 and 9)

QUESTION

Calculate the weight of zinc and hydrochloric acid needed to produce enough hydrogen to fill a spherical balloon 5 meters in diameter (16.4 feet), if the pressure is one atmosphere and the temperature is 298°K. (Volume of a sphere is $4/3\pi r^3$.) The reaction is $Zn + 2HCl \rightarrow ZnCl_2 + H_2$.

PROBLEM

Calculate the equivalent weight of zinc from Cavendish's data, if the flammable air (hydrogen) was collected over water. Assume that atmospheric pressure was near 760 torr, and that the day was fairly warm in England, maybe 25°C. Take into account the vapor pressure of water, and subtract the weight of water in the gas sample. The weight of water can be obtained by multiplying the fraction by weight of water in the gas by the weight of the gas.

EXPERIMENT 12

Equipment and supplies

One 200-mm test tube

Two two-hole rubber stoppers, No. 6

15 Inches soft glass tubing

10 Inches hard glass tubing (Pyrex)

Rubber or plastic tubing

4 Inches wire

One 13- × 100-mm test tube

One drying tube

One burner

One wing tip (flame spreader)

Clamp and ring stand

One cloth towel

One 250-ml Erlenmeyer flask, or bottle

One funnel

Analytical balance

0.4 g Unknown oxide or known oxide

10 g Mossy zinc

30 g Anhydrous calcium chloride or calcium sulfate

50 ml $6N$ Sulfuric acid

2 ml $1M$ Copper sulfate

10 ml Concentrated nitric acid

Glycerol

Time requirement

Two hours

THE COMBINING WEIGHT OF A METAL BY REDUCTION OF THE OXIDE

The object of this experiment is to determine the equivalent weight of a metal by reduction of the oxide, using hydrogen as the reducing agent. The equivalent weight of an element is the weight that will combine with, or displace, 1 g of hydrogen or 8 g of oxygen.

PROCEDURE

Caution: Throughout this experiment, remember that hydrogen mixed with air forms a powerful explosive, so do not bring a flame near the apparatus until you are sure all air has been displaced.

Set up the apparatus that is shown in Figure 12-1. If you need to make the glass pieces, read Section 0.7 on glassworking. Also, be careful to follow safety instructions in Sections 0.2 and 0.7 concerning inserting glass tubes into rubber stoppers. Avoid injury! Notice that hydrogen is formed in a bottle by reaction of zinc and dilute sulfuric acid. The hydrogen flows through a calcium chloride drying tube and then over

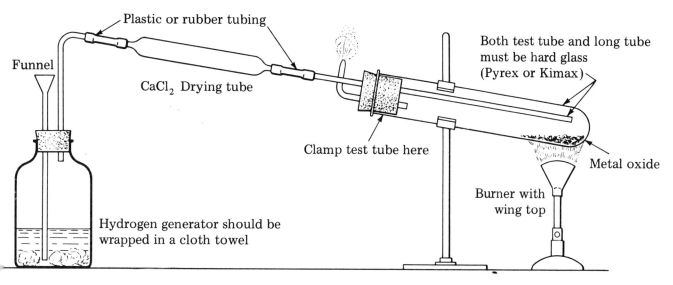

Figure 12-1. Apparatus in operation.

the unknown metal oxide. The hydrogen finally escapes at the little jet. The hole in the jet must not be too small. A diameter of 1/16 inch is satisfactory. Calcium chloride strongly absorbs water vapor, thus it dries the hydrogen. Clean and dry (by heating) a large (170 or 200 mm) hard glass test tube. Let it cool, then weigh it. The test tube may be suspended by a wire from the hook above the balance pan. Add about 0.3-0.5 g of the unknown oxide to the test tube, then weigh the tube and oxide. Record the weights of the test tube, the test tube plus oxide, and the oxide, obtained by difference.

Jiggle the test tube so that the oxide is spread out in a thin layer at the bottom of the test tube. Connect the test tube to the apparatus without touching the glass tubing to the oxide. Place some zinc (15-20 g) in the hydrogen generator. Be sure all connections are airtight. Ask the instructor to inspect the apparatus.

Pour about 50 ml of dilute H_2SO_4 into the hydrogen generator. If evolution of hydrogen is very slow, it may be speeded up by adding 2 ml of copper sulfate solution. Permit evolution of hydrogen to continue for about two minutes. Then test for complete displacement of air in the following way. Hold a small test tube inverted over the jet for about 15 seconds. Then quickly hold the mouth of the test tube to a flame situated at least two feet from any part of the rest of the apparatus. If the hydrogen burns quietly, the apparatus is safe. If it pops, except gently, you must wait a few moments, then repeat the test. Finally, you may hold the test tube of burning hydrogen near the jet to ignite the hydrogen coming out. Under no circumstance should a lighted match, Bunsen burner, or other flame be used for this purpose. As soon as you are satisfied that all air has been displaced, heat the test tube strongly, using the wing tip (flame spreader) on the Bunsen burner. Continue heating for about five minutes. The general reaction can be written

$$n H_2 + MO_n \rightarrow M + n H_2O$$

Record changes in the oxide, and any other observations you make as to substances formed. Try to identify, by its appearance, any substance formed in the cooler part of the test tube. Do not smell the escaping hydrogen. After the reduction is complete, heat the rest of the test tube slightly to drive off the condensed water. Do not heat strongly enough to soften or melt the rubber stopper.

After about five minutes, allow the test tube to cool, but be sure that hydrogen is still being evolved. It may be wise to consult the instructor concerning completeness of reduction. When the test tube is quite cool, remove and weigh it. Record the weight of test tube plus metal, and of metal alone. Compute the equivalent weight of the metal.

The test tube can be cleaned with nitric acid.

The reduction can be performed with the oxide contained in a porcelain boat. However, the amount of heat necessary to heat the boat to the required temperature is not easy to attain with the ordinary burner, especially without softening the glass tube in which the boat is placed.

SUPPLEMENTARY READING

Dickerson, Gray, and Haight (Chapter 1, Sections 4 and 9, Chapter 3, Section 1)

QUESTIONS

1. What effect would incomplete conversion of the metal oxide to metal have on the combining weight determined?
2. The calcium chloride drying tube removes water vapor from the hydrogen gas before it enters the reaction tube. Liquid water, however, may be observed in the cooler part of the test tube during the reduction. How is this water formed?

PROBLEMS

1. In 1900, Richards and Baxter (Harvard University)[30] used a method similar to this experiment to determine the atomic weight of iron, obtaining 55.86 instead of the earlier accepted value of 56.00. Suppose that 0.5000 g of ferric oxide (Fe_2O_3) upon reduction with hydrogen gave 0.3498 g of iron metal. Show the reasoning that leads to the atomic weight of iron.

2. Atomic weights can be obtained accurately from mass spectroscopic measurements. From these measurements, the relative abundance of the various iron isotopes can be obtained, as well as their exact atomic masses. Calculate the atomic weight of iron from the following abundances and masses, which are all relative to ^{12}C = 12.0000 amu (atomic mass units):

^{54}Fe, 5.82%; 53.9396 amu

^{56}Fe, 91.66%; 55.9349 amu

^{57}Fe, 2.19%; 56.9354 amu

^{58}Fe, 0.33%; 57.9333 amu

Carry out the calculation to six significant figures. In 1953, the following abundances and masses were the accepted values. These measurements were relative to ^{16}O = 16.0000 amu. Calculate the atomic weight from these data also. Did the atomic weight change (that is, did the accepted value change)?

^{54}Fe, 5.90%; 53.95766 amu

^{56}Fe, 91.52%; 55.95332 amu

^{57}Fe, 2.245%; 56.95477 amu

^{58}Fe, 0.33%; 57.95083 amu

Because ^{16}O is 15.9994 amu on the scale in which ^{12}C is 12.0000 amu, to convert from the ^{16}O scale to the ^{12}C scale multiply by 0.99996.

EXPERIMENT 13

Equipment and supplies

One 500-ml Erlenmeyer flask

One 50-ml burette

Two 250-ml Erlenmeyer flasks

Analytical balance

One pipette (if unknown is a solution)

Burette brush (available in laboratory)

Burner, wire gauze, ring stand

1 g Unknown acid

2 g Oxalic acid

15 ml 6M Sodium hydroxide

1 ml Phenolphthalein indicator solution

Time requirement

Two hours

VOLUMETRIC ANALYSIS: TITRATION OF ACIDS AND BASES

The purpose of this experiment is to determine the concentration of an acid, and to determine the equivalent weight of an unknown acid.

INTRODUCTION

The concentration of a known acid can be determined by finding the volume of the acid that exactly neutralizes a definite amount of a base. The neutral point, or end point, of the reaction is found by use of an indicator. Indicators are substances that sharply change color as the solution changes from acid to base, or vice versa. The use of the burette for titration, that is, for measuring volumes of titrant, is described in Section 0.5. The use of pipettes also is described there.

PROCEDURE

Part 1: Standardization of the sodium hydroxide titrant

A solution of approximately $6M$ NaOH will be available in the laboratory. This must be diluted to make an approximately $0.2M$ solution. The volume that is needed depends on whether both Parts 2 and 3 are to be done. If only Part 1 and either Parts 2 or 3 are to be done, about 250 ml of $0.2M$ NaOH will be needed. About 400 ml should be plenty for Parts 1, 2, and 3.

The water used to make the dilute solution must be free of carbon dioxide, because carbon dioxide reacts with sodium hydroxide

$$NaOH + CO_2 \rightarrow NaHCO_3$$

Consequently, boil about 300 ml of distilled water in a clean Erlenmeyer flask for a few minutes to outgas the water. Let the flask cool, while it is covered with aluminum foil or a loosely fitting cork. When the water is at room temperature, mix the calculated volume of $6M$ NaOH needed with a calculated volume of water to make the 250 ml of $0.2M$ NaOH. Keep the flask tightly corked, especially after the NaOH has been standardized.

Oxalic acid, $H_2C_2O_4 \cdot 2H_2O$, is suitable for standardizing sodium hydroxide, because it is easily obtained in pure form and is readily weighed. Notice that oxalic acid has two replaceable hydrogen atoms, and that the molecule has two molecules of water of crystallization. Calculate the weight of $H_2C_2O_4 \cdot 2H_2O$ needed for a titration volume of 30-50 ml. The larger the titration volume (less than one burette full), the smaller is the percentage of error, because the error is mostly in reading the burette. Weigh out two samples of approximately the correct amount of oxalic acid dihydrate on two pieces of smooth paper, using a triple beam balance. Then weigh an Erlenmeyer flask (125 ml) accurately on an analytical balance, add one oxalic acid sample, and reweigh accurately. Do the same with the second sample. Remember to number the flasks. Add 25 ml of water to each flask to dissolve the oxalic acid.

Rinse the 50-ml burette three times with distilled water, and then twice with about 10 ml of the $0.2M$ sodium hydroxide solution. Clamp the burette in position, then fill the burette to just above the zero mark with the $0.2M$ NaOH. Be sure to let any air bubbles out of the tip of the burette. Bring the sodium hydroxide solution meniscus down to the zero mark. Read and record the burette reading to ±0.02 ml. The burette can be read more accurately if a card with the bottom half blackened is held directly behind the meniscus, so that the meniscus appears just to touch the black background. (See Figure 0-15.)

Add three drops of phenolphthalein indicator solution to each flask. Set one flask under the burette, and add sodium hydroxide dropwise. Swirl the flask constantly. Stop as soon as the first permanent pale pink color appears. This is the end point. Before the end point is reached, a temporary pink color will be seen where the NaOH drop hits, but it will disappear when the flask is swirled. Read the burette to ±0.02 ml and record the reading. Repeat the titration with flask No. 2.

From the weight of oxalic acid dihydrate and the volume of sodium hydroxide used in each sample, calculate the exact number of moles of sodium hydroxide. Calculate the molar concentration of the titrant. The results of the duplicate samples should be within 1% of each other. If they are not, repeat the determination with two more samples of oxalic acid dihydrate. Record the average concentration of the standardized base.

Part 2: Determination of the concentration of an acid solution

Obtain an acid solution of unknown concentration. Clean a 5-ml pipette, and rinse it three times with distilled water. Rinse it twice with small amounts of the acid solution. *Note:* Do not use your mouth to suck up the solution! Use a pipette bulb! (See Section 0.5.) Pipette a 5-ml aliquot into each of two 125-ml Erlenmeyer flasks. Add three drops of phenolphthalein solution to each flask, and titrate each solution as before. From the known volume of unknown acid solution (5.00 ml) and the titration volumes, calculate the concentration of the acid solution, if the identity of the acid is known. If the identity of the acid is not known, only the normality of the acid solution can be calculated. A 1.00-normal acid solution (1.00N) is a solution having one mole of hydronium ions, or acidic hydrogens (which is one equivalent), per liter of solution.

Part 3: Determination of the equivalent weight of an acid

An equivalent of an acid is the weight that just neutralizes one mole of hydroxide ions. Obtain an unknown acid from the instructor. Weigh two samples of the unknown in two Erlenmeyer flasks. The samples should be approximately 0.5 g in weight. Weigh the flask empty, then add the unknown and reweigh accurately. Add 25 ml of distilled water and three drops of phenolphthalein indicator solution to each flask. Titrate each to the first permanent pink color using 0.2M sodium hydroxide standardized as described in Part 1. The total titer (the number of milliliters of sodium hydroxide used) should be less than the volume of the burette (50 ml), but not too small. A titer of 20 ml probably will give sufficient accuracy. If the titer is outside these limits, it would be advisable to weigh samples of a size that will require a titer within 20-50 ml.

Calculate the equivalent weight of the acid. Estimate the errors in your result, assuming weighing to ±0.0005 g and titration readings of ±0.05 ml.

SUPPLEMENTARY READING

Dickerson, Gray, and Haight (Chapter 4)

Masterton and Slowinski (Chapter 18)

Sienko and Plane (Chapter 9)

Mahan (Chapter 6, Section 7)

Brescia, *et al.* (Chapters 17 and 21)

Brown (Chapter 12, Section 6)

Pauling (Chapter 19)

Hammond, Osteryoung, Crawford, and Gray (Chapter 9, Section 3)

QUESTIONS

1. How is the result (the equivalent weight of an unknown acid, for example) affected if you titrate past the end point? What would the percentage of error be if you titrated one drop past the end point, and the titration volume was about 30 ml?

2. What would be the effect on the result if the burette were dirty so that droplets of titrant (NaOH solution, for example) adhered to the walls of the burette after the solution was titrated?

3. What would happen if the sodium hydroxide titrant had been exposed to carbon dioxide in the air for a long time after standardization? Before standardization?

4. Why does the amount of water (say, 25 ml versus 35 ml) used to dissolve the unknown acid samples not matter in the titrations?

5. What might be the effects of not rinsing the burette with the sodium hydroxide titrant before filling the burette for a titration? (For example, the burette might be wet from cleaning with distilled water.)

PROBLEMS

1. Assume reasonable error limits in your measurements of weight and volume. Then calculate the maximum error in your determinations.

2. Give exact instructions describing the preparation of a 1.00M solution of hydrochloric acid in water from the stock laboratory solution that is 6.00M in HCl. You may assume that volumes of the HCl and water are additive. However, how could you make the dilution exactly without assuming that the volumes are additive?

3. What volume of 1.00M hydrochloric acid is required to neutralize 25.0 ml of 0.065M sodium hydroxide?

4. What volume of this 1.00M hydrochloric acid is required to neutralize 1.00 g of calcium hydroxide, $Ca(OH)_2$?

EXPERIMENT 14

Equipment and supplies

One 500-ml flask

Two 250-ml Erlenmeyer flasks

Burner, wire gauze, ring stand

One 50-ml burette

1.6 g Potassium permanganate

1 g Sodium oxalate

120 ml 3M (6N) Sulfuric acid

25 ml Unknown hydrogen peroxide solution (Part 2)

1.5 g Unknown oxalate (Part 3)

Time requirement

Part 1 and either Parts 2 or 3: two hours

Parts 1, 2, and 3: two and one-half hours

VOLUMETRIC ANALYSIS: POTASSIUM PERMANGANATE

In this experiment, the concentration of aqueous hydrogen peroxide and the percentage of oxalate in an unknown oxalate sample will be determined by titration with potassium permanganate.

PROCEDURE

Part 1: Preparation and standardization of 0.2N potassium permanganate

Potassium permanganate is a strong oxidizing agent. Permanganate itself will cause purple spots. However, since it is an oxidizing agent, it is reduced to manganese dioxide, which is brown. Therefore, to prevent getting brown spots on clothes, skin, and so forth, handle the solid and solution carefully.

Weigh 1.6 g of potassium permanganate on a piece of smooth paper on a triple beam balance. Add the 1.6-g sample to 100 ml of boiling distilled water and stir until the sample dissolves. After the crystals have dissolved, add the solution to 150 ml of distilled water in a 500-ml Erlenmeyer flask, and mix the solution well. Let the solution cool to room temperature, or cool the flask in a pan of cold water. Keep the flask tightly closed with a rubber stopper. The concentration of this solution is approximately 0.04M, or 0.2N.

Solutions of potassium permanganate made in this way do not keep well because the impurities, especially manganese dioxide, catalyze the decomposition of permanganate. Consequently, the permanganate solution must be standardized on the day it is used. If the permanganate solution is boiled and filtered through a sintered glass funnel, and if it is protected from light and contamination, the concentration will remain constant for weeks. The permanganate solution may be standardized conveniently by titrating sodium oxalate in acidic solution. The products of the reaction are carbon dioxide and manganous ion

$$5C_2O_4^{2-} + 2MnO_4^- + 16H^+ \rightarrow 10CO_2 + 2Mn^{2+} + 8H_2O$$

Weigh two samples of approximately 0.5 g of sodium oxalate (which has been dried at 105°C) on smooth paper. Then weigh two 250-ml Erlenmeyer flasks accurately. Add the sodium oxalate and reweigh accurately. Do the same for the other flask. Add 120 ml of water and 20 ml of 6N sulfuric acid. Heat the solution of sodium oxalate to 80°C-90°C (do not boil), and titrate slowly with permanganate to the first permanent pink color, swirling the flask constantly. Add the permanganate in portions, letting the color disappear before adding more. The temperature must be above 60°C at the end point. To minimize loss of heat, set the flask on two paper towels for insulation during the titration. Notice that the first drops of permanganate react slowly with the oxalate, but later drops react almost instantly. Manganous ion, which is a product, catalyzes the reaction.

From the weight of sodium oxalate, the volume of permanganate, and the mole relationships in the equation, calculate the exact normality of the potassium permanganate solution.

Part 2: Determination of hydrogen peroxide concentration

Obtain a sample of hydrogen peroxide solution. Using a pipette bulb, pipette a 5-ml aliquot of the solution into each of two 250-ml Erlenmeyer flasks. Add 50 ml of water and 20 ml of 6N sulfuric acid to each flask. Titrate the solution in each flask with the standard permanganate solution to the final faint pink color.

The reaction is represented by the following reactants and products in acidic solution:

$$MnO_4^- + H_2O_2 \rightarrow O_2 + Mn^{2+} + H_2O$$

Balance the equation and calculate the concentration of the hydrogen peroxide solution.

Part 3: Determination of the percentage of oxalate

Obtain an oxalate sample. Weigh about 0.5 g of the sample on a smooth piece of paper. Weigh an Erlenmeyer flask accurately and reweigh accurately after adding the half-gram sample of oxalate. Add 120 ml of water and 20 ml of 6N sulfuric acid to the flask, and titrate with standardized permanganate solution. The first permanent faint pink is the end point. The titration volume should be between 20 and 50 ml. If it is not, adjust the size of the sample accordingly. In any case, do duplicate determinations. From the relations of the balanced equation in Part 1, the weight of oxalate, and the titration volume, calculate the weight of oxalate ion present in the sample. Calculate also the percentage by weight of oxalate ion in the sample.

SUPPLEMENTARY READING

Dickerson, Gray, and Haight (Chapter 6, Sections 4 and 5)

Masterton and Slowinski (Chapter 15, Section 1, and Chapter 20, Section 4)

Sienko and Plane (Chapter 4)

Mahan (Chapter 7, Sections 1-3 and 6)

Brescia, *et al.* (Chapter 21)

Brown (Chapter 21, Section 7, and Appendix A)

Pauling (Chapter 11, Section 4)

Hammond, Osteryoung, Crawford, and Gray (Chapter 11, Section 6)

QUESTIONS

1. What is the change in oxidation state for manganese in $KMnO_4$ in this reaction, which takes place in acidic solution?

2. What is the normality of a $1.0M$ solution of $KMnO_4$ when used as an oxidizing agent in acidic solution?

3. Why must the oxygen atoms in hydrogen peroxide be oxidized (an increase in oxidation state) in the reaction with permanganate ion that forms manganous ion? In other words, why cannot both manganese and oxygen decrease (or increase) in oxidation state?

PROBLEMS

1. In the analysis of steel, the manganese content can be determined by dissolving a 1.00-g sample of the steel in dilute nitric acid and proceeding as follows. The manganous ion formed in dissolving the steel is oxidized to permanganate ion (purple) by adding sodium bismuthate

$$2Mn^{2+} + 5NaBiO_3 + 14H^+ \rightarrow 2MnO_4^- + 5Na^+ + 5Bi^{3+} + 7H_2O$$

The excess of sodium bismuthate is filtered off. Then the solution of MnO_4^- is titrated with a standard $0.025M$ solution of ferrous ion in acidic solution until the purple color of the permanganate ion disappears. In a certain analysis, 15.0 ml of the standard ferrous ammonium sulfate solution was required to titrate the permanganate. Balance the following equation and compute the percentage of manganese in the steel:

$$Fe^{2+} + MnO_4^- \rightarrow Mn^{2+} + Fe^{3+}$$

2. A practical analytical procedure for determining calcium involves (1) dissolving the sample containing calcium, (2) precipitation of the calcium as $CaC_2O_4 \cdot H_2O$ by adding ammonium oxalate, and (3) titration of the oxalate precipitate with permanganate. A 0.500-g sample of limestone (containing calcium carbonate as one constituent) was treated as outlined, yielding a precipitate of calcium oxalate. The calcium oxalate hydrate was filtered off and then dissolved in dilute sulfuric acid and titrated with $0.200N$ $KMnO_4$ solution. Exactly 31.50 ml was required in the titration. Calculate the percentage of calcium carbonate in the limestone.

EXPERIMENT 15

Equipment and supplies

One 150-mm test tube
One two-hole rubber stopper, No. 2
Glass tubing
Rubber or plastic tubing
Vacuum line or aspirator
One filter flask with one-hole rubber stopper
One Hirsch funnel with rubber fitting
One 100-ml beaker
Burner, wire gauze, ring stand
One watch glass
Triple beam balance
One 200-mm test tube
One two-hole rubber stopper, No. 6
Glass tubing
Rubber or plastic tubing
One evaporating dish
One 250-ml beaker
One 50-ml burette
Analytical balance
One 10-ml graduated cylinder
One 100-ml graduated cylinder
Two 125-ml Erlenmeyer flasks
One sintered glass filter crucible
Spectrometer

2.5 g Cobaltous chloride hexahydrate
0.7 g Ammonium chloride
5 ml Concentrated ammonia
0.2 g Activated carbon
Filter paper for Hirsch funnel
Filter paper, 11 or 15 cm
17 ml Concentrated hydrochloric acid
Ice
5 ml 95% Ethanol
7 g Ammonium carbonate
13 ml Concentrated ammonia
2.5 g Cobaltous nitrate hexahydrate
3.3 g Mercuric nitrate
1 ml 10% Sodium nitroprusside indicator
1 ml 6M Nitric acid
1 g Potassium chloride
10 ml 1M Silver nitrate
Glycerol
5 ml 10% Ethylenediamine solution
15 ml Methyl alcohol

Time requirement

The oxidation in Parts 1 and 2 takes three hours. Parts 1 and 2 can be done at the same time. For efficiency, other experiments should be started or finished during aeration; for example, Experiments 26 or 18. After oxidation, workup for Parts 1 and 2 takes about two hours, and Part 3, two hours. Part 4 takes two hours, and Part 5 one and one-half hours.

COMPLEX IONS OF COBALT

The purpose of this experiment is to prepare hexaamminecobalt(III) chloride and chloropentaamminecobalt(III) chloride. These compounds will then be analyzed to determine the position of the chloride ion. *Cis*- and *trans*-dichlorobisethylenediaminecobalt(III) chlorides also will be prepared and analyzed. The isomerization of cis to trans will be studied.

INTRODUCTION

In 1893, Alfred Werner[31] first correctly formulated a theory of the structure of coordination compounds. Before this time, they were classed as "complex compounds," since they were written as though made up of more than one compound; for example, $CoCl_3 \cdot 6NH_3$, a complex compound of cobaltic chloride and ammonia.

Following are some of the reactions of the yellow luteocobaltic chloride.

$$CoCl_3 \cdot 6NH_3 + HCl \; (concd) \xrightarrow{100°C} \text{no reaction}$$

$$CoCl_3 \cdot 6NH_3 + 3AgNO_3 \; (aq) \xrightarrow{cold} Co(NO_3)_3 \cdot 6NH_3 + 3AgCl$$

$$CoCl_3 \cdot 6NH_3 + \frac{3}{2}H_2SO_4 \; (concd) \longrightarrow \frac{1}{2}[Co_2(SO_4)_3 \cdot 12NH_3] + 3HCl$$

$$CoCl_3 \cdot 6NH_3 + 3KOH \xrightarrow{100°C} \frac{1}{2}(Co_2O_3) + 6NH_3 + 3KCl + \frac{3}{2}H_2O$$

$$CoCl_3 \cdot 6NH_3 + \frac{3}{2}Ag_2O + \frac{3}{2}H_2O \xrightarrow{moist} Co(OH)_3 \cdot 6NH_3 + 3AgCl$$

$$CoCl_3 \cdot 6NH_3 + 3AgX \longrightarrow CoX_3 \cdot 6NH_3 + 3AgCl$$

$$Co(OH)_3 \cdot 6NH_3 + 3HX \longrightarrow CoX_3 \cdot 6NH_3 + 3H_2O$$

From these reactions, Werner noticed that six ammonia molecules always remain in the molecule, and fail to react even with concentrated hydrochloric acid. Therefore, he concluded that the six ammonia molecules are tightly bound to the cobalt atom. The fact that cold silver nitrate precipitates three chloride ions was taken as evidence that these three chlorides were attached to cobalt by ionic bonds, as chloride is

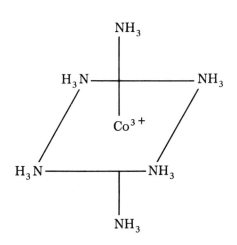

Figure 15-1. Conventional drawing of the hexaamminecobalt(III) chloride complex, representing its geometry.

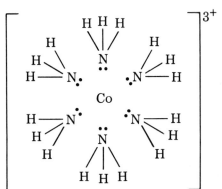

Figure 15-2. Lewis structure of hexaamminecobalt(III) chloride.

"attached" in sodium chloride. The equivalent conductance, Λ, was appropriate for four ions. The Van't Hoff coefficient, i, from freezing point lowering or osmotic pressure measurements was 3.9-4.2, which is appropriate for four ions.

Simple cobaltous (Co^{2+}) salts are stable to oxidation. The simple cobaltic (Co^{3+}) salts are not stable, and are so reactive that they oxidize water to oxygen. However, cobaltous complex ions can be oxidized to the cobaltic complex ions, which are not easily reduced. Cobalt has 27 electrons, and the lowest energy electronic configuration is $(Ar)3s^2 3p^6 4s^2 3d^7$. The cobaltic ion has three electrons less, and is 12 electrons short of the inert gas configuration of krypton $[(Ar)3s^2 3p^6 3d^{10} 4s^2 4p^6]$. By sharing six pairs of electrons from donating ions or molecules, six coordinate covalent bonds can be formed. All six coordinate cobalt compounds are octahedral.

An octahedral arrangement is one in which the ligands are situated at the corners of an octahedron, at the center of which is the metal atom. The structure of the so-called luteocobaltic chloride is conventionally represented as in Figure 15-1. Each ammonia ligand is equivalent, although the conventional drawing tends to imply that two are different. A Lewis (electron dot) structure for the same complex is shown in Figure 15-2. It shows that the unshared pair of electrons on nitrogen in ammonia is the pair of electrons used to bond to cobalt. This Lewis structure is not intended to represent the geometry of the complex, as is Figure 15-1. The correct systematic name for this complex is *hexaamminecobalt(III) chloride*. The name luteocobaltic chloride is from an old system (*luteo* means yellow).

Cobalt complexes occur in nature in various forms of vitamin B-12. Cobalt(III) in these structures is coordinated to four nitrogen atoms of the corrin ring system (Figure 15-3), with additional ligands above (wedge-shaped bond) and below (dashed bond) the plane of the corrin ring (Figure 15-4). Other metal complexes such as chlorophyll (Mg^{2+}), and hemoglobin (Fe^{2+}) are important in living systems also. Structures of these will be found in most textbooks. Many other metals are known to be necessary for life in minute amounts (e.g., Cu^{2+}, Mn^{2+}), but their functions are not understood.

A large number of cobalt complexes have been prepared. Typical preparative sequences, like those in this assignment, involve oxidation of cobaltous ion in the presence of an appropriate ligand. A ligand can be displaced by another ligand also. In Parts 1 and 2, hexaamminecobalt(III) chloride and chloropentaamminecobalt(III) chloride are prepared. Analytical procedures are described in Part 3 for the determination of ionic chloride.

In Part 4, preparative procedures are given for *cis*- and *trans*-dichlorobisethylenediamine cobalt(III) chloride, adapted from an article in the *Journal of Chemical Education* by R. D. Foust and P. C. Ford.[32] These compounds are isomers, differing only in that the trans isomer has the chlorides on opposite sides of the cobalt, whereas in the cis isomer, the chlorides are adjacent to each other. These cations are shown in Figure 15-5, in which two conventions for drawing them are

Introduction 109

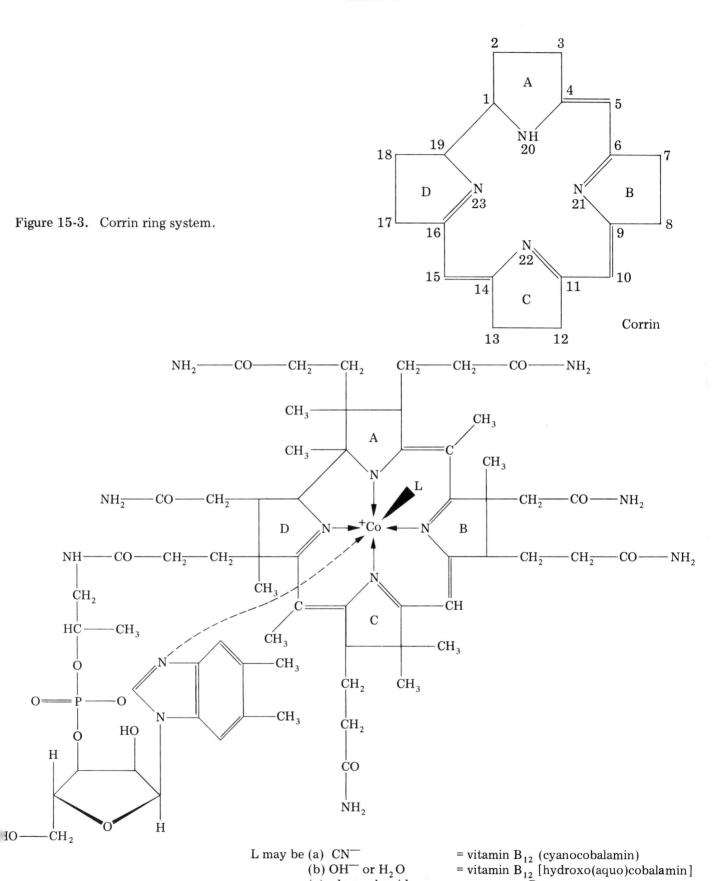

Figure 15-3. Corrin ring system.

Figure 15-4. Vitamin B_{12} structure.

L may be (a) CN^- = vitamin B_{12} (cyanocobalamin)
(b) OH^- or H_2O = vitamin B_{12} [hydroxo(aquo)cobalamin]
(c) adenosyl residue = coenzyme B_{12}
(d) alkyl groups = coenzyme analogs

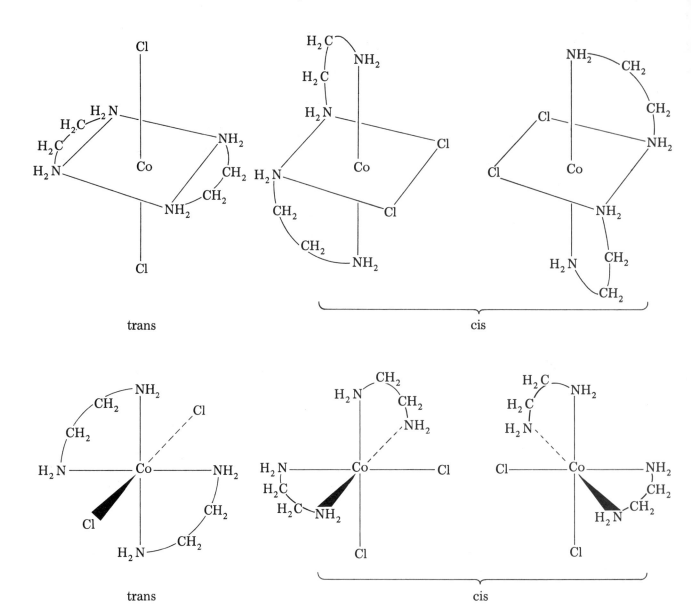

Figure 15-5. Two ways of drawing *cis*- and *trans*-dichlorobisethylenediaminecobalt(III) ions; each ion has a +1 charge.

used. In the lower structures, all solid bond lines to cobalt are understood to lie in the plane of the paper. The wedge-shaped bond comes out toward the reader, and the dashed bond line extends behind the plane of the paper. The ethylenediamine ligand is attached to cobalt at both nitrogen ends. Ethylenediamine could be regarded as two ammonia molecules joined by a carbon chain. The cis isomer (Figure 15-5) has two isomeric forms that differ in that they are mirror images of one another and cannot be superimposed on one another. Your hands have such a relationship to one another. The geometrical isomers, cis and trans, have different physical properties that make it possible to separate them. The two cis isomers, however, have the same physical properties, except that one rotates plane-polarized light counterclockwise (to the left), and the other clockwise (to the right). Isomers that differ only in that they are mirror images (and nonsuperimposable, so they are not identical) are called optical isomers. Because their physical properties are the same, they cannot be separated from one another as the chloride salt. The cis isomer formed is a 50/50 mixture of the two optical isomers; consequently, the cis isomer does not rotate plane-polarized light.

PROCEDURE

Part 1: Preparation of hexaamminecobalt(III) chloride

In a 150-mm test tube, mix 1.0 g (0.0042 mole) of cobaltous chloride hexahydrate, $CoCl_2 \cdot 6H_2O$, and 0.70 g of ammonium chloride with 1 ml of water until most of the salts are dissolved. Add 3 ml of concentrated ammonia and half a spatula of activated carbon (fresh from the package). Bubble air through the solution for at least 1 1/2 hours, by connecting the aspirator shown in Figure 15-6. Bubble air for three hours if the carbon catalyst is not absolutely fresh. When the aspirator is used for suction, a trap should be placed between the aspirator and the reaction tube to prevent water from backing up from the aspirator. Compressed air can be used, but it is usually difficult to regulate. During the bubbling of air, the color of the solution will change from reddish to yellowish brown, as the result of the oxidation of Co(II) to Co(III) by oxygen. The carbon acts as a catalyst. Stopper the test tube until the next laboratory period. Be sure to return special apparatus to the storeroom.

Filter the reaction mixture with suction. The filter cake contains the product along with carbon. Stir the filter cake with 6-8 ml of water

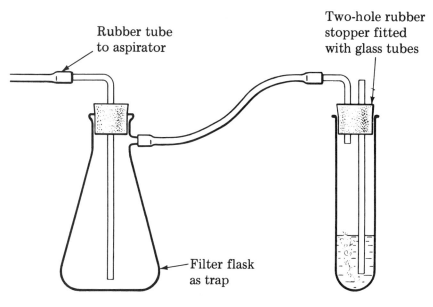

Figure 15-6. Reaction tube connected to an aspirator.

in a small beaker to which three drops of concentrated HCl has been added. The solution should now be acidic to litmus paper. If the solution is not acidic, add a drop of concentrated HCl to the solution. Repeat if necessary. Heat the solution to 50°C-60°C and filter it while hot, using suction. The product is in the solution. Add 1.5 ml of concentrated HCl to the hot solution to precipitate the product. The HCl reduces the solubility of the chloride salt by the common ion effect. The hot solution should be allowed to cool to room temperature slowly, then it should be cooled in ice water. Filter off the yellow crystals with suction. Wash with 2 ml of 95% ethyl alcohol. Dry on a watch glass, then weigh the dried product. Calculate the percentage of the theoretical yield that you obtain. Analyze the product for ionic chloride by one of the methods in Part 3.

Part 2: Preparation of chloropentaamminecobalt(III) chloride

Add 5.0 g of ammonium carbonate to a large test tube (200-mm). Add 25 ml of water to dissolve the solid. Then add 13 ml of concentrated ammonia. Dissolve 2.5 g (0.0086 mole) of cobaltous nitrate hexahydrate, $Co(NO_3)_2 \cdot 6H_2O$, in 5 ml of water. Now add this second solution to the first solution and mix. Bubble air through the combined solutions for three hours, using a setup similar to that shown in Figure 15-6. The color of the solution will change from blue (cobaltous) to red, then to deep red (cobaltic). The reaction solution may be stoppered until the next laboratory period.

Pour the solution into an evaporating dish. Evaporate the solution to 10 ml by heating the dish on a beaker of boiling water. Periodically add 1.5 g of powdered ammonium carbonate in small amounts. Let the solution cool to cause crystallization. Cool the dish well (15 to 30 minutes) in ice water, and then with suction filter off the purple-red crystals of carbonatotetraamminecobalt(III) nitrate, $[Co(NH_3)_4CO_3]NO_3 \cdot 1/2H_2O$. Weigh your product and calculate the percentage yield.

In the following procedure, use all (or almost all) of the product you have just obtained. Dissolve the carbonatotetraamminecobalt(III) nitrate in 8-10 ml of water in a large test tube. Add concentrated hydrochloric acid until all the carbon dioxide is expelled (bubbles stop). About 1 ml should be enough for the complete reaction

$$[CoCO_3(NH_3)_4]^+ + 2H_3O^+ \rightarrow H_2CO_3 + [Co(NH_3)_4(H_2O)_2]^{3+}$$

$$H_2CO_3 \rightarrow H_2O + CO_2\uparrow$$

Neutralize the solution with concentrated ammonia and add 1 ml excess. Heat the test tube containing the solution in a beaker of boiling water for 20 minutes. The reaction is

$$[Co(NH_3)_4(H_2O)_2]^{3+} + NH_3 \rightarrow [Co(NH_3)_5H_2O]^{3+} + H_2O$$

Cool the mixture, then add 10 ml of concentrated hydrochloric acid. Heat the test tube in the water bath for another half hour. The following reaction occurs:

$$[Co(NH_3)_5H_2O]^{3+} + Cl^- \xrightarrow{H^+} [Co(NH_3)_5Cl]^{2+} + H_2O$$

Cool the test tube in ice water. Filter off the violet crystals, wash them with a few drops of ice water, and then with alcohol. Weigh the dry product, which is purpureocobaltic chloride (the old name) or chloropentaamminecobalt(III) chloride, $[Co(NH_3)_5Cl]Cl_2$. Calculate the percentage yield. Analyze the product for ionic chloride, and for covalently bound chloride, by the procedures in Part 3.

Part 3: Analysis for chloride ion

Mercuric nitrate titration

The ionic chlorides in complex salts can be titrated with mercuric nitrate to form mercuric chloride, which, though water soluble, is only slightly dissociated

$$2Cl^- + Hg^{2+} \rightarrow HgCl_2$$

To detect the point at which mercuric ion is in excess of the chloride ion, an indicator, sodium nitroprusside (sodium nitroferricyanide), $Na_2[FeNO(CN)_5] \cdot 2H_2O$, is added. Mercuric nitroprusside is very soluble; consequently, when the chloride is consumed upon addition of mercuric nitrate, the excess mercuric ion forms mercuric nitroprusside, which makes the solution turbid. This cloudiness is the end point of the titration. The reaction is

$$Hg^{2+} + [FeNO(CN)_5]^{2-} \rightarrow Hg[FeNO(CN)_5]$$

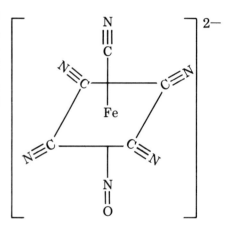

Figure 15-7. Nitroprusside ion.

The structure of the nitroprusside ion is shown in Figure 15-7.

Both mercuric and nitroprusside compounds are POISONOUS! Wash your hands after using them.

Prepare the 0.1M mercuric nitrate titrant by adding approximately 3.3 g of the salt $Hg(NO_3)_2$ to distilled water, containing 1 ml of 6M nitric acid, and making the solution up to exactly 100 ml. Weigh the salt in a flask on an analytical balance to the nearest 5 mg. The mercuric nitrate is very hygroscopic, so keep the reagent bottle closed. If the mercuric nitrate is not anhydrous reagent grade, the solution of mercuric nitrate must be standardized against weighed amounts of potassium chloride.

If the mercuric nitrate is anhydrous, the titrant made from it will be accurate enough for the purposes of this experiment, making this standardization unnecessary. To standardize the mercuric nitrate titrant, weigh two samples of potassium chloride, each weighing about a half gram, into two Erlenmeyer flasks. Dissolve each sample in 50 ml of distilled water and add three drops of 10% sodium nitroprusside to each. Titrate each sample with mercuric nitrate. Calculate the exact concentration of the titrant.

Weigh a sample of approximately 0.50 g of your salt accurately, using an analytical balance. Dissolve this sample in exactly 50.00 ml of distilled water. Titrate 10 ml of this solution with the mercuric nitrate titrant after adding three drops of 10% sodium nitroprusside and 40 ml of distilled water. Calculate the equivalent weight of the salt. Is it consistent with the formula given?

Silver chloride precipitation

Silver nitrate may be used to precipitate chloride ion as silver chloride

$$Ag^+ + Cl^- \rightarrow AgCl$$

The solution should be slightly acidic (with nitric acid) or silver hydroxide may be obtained. Silver chloride is a very slightly soluble white solid. It may be dried in an oven at 120°C. However, it is light sensitive, yielding silver and chlorine

$$2AgCl \overset{light}{\rightarrow} 2Ag + Cl_2$$

Add an excess of silver nitrate solution to a solution of a little of either salt that has just been acidified with dilute nitric acid. Centrifuge and separate the precipitate. Make the solution strongly acidic with nitric acid and heat it in a beaker of boiling water. The solution containing purpureocobaltic chloride will develop a second precipitate. The hexaamminecobaltic chloride will not. Why?

This procedure can be made quantitative. To do so, use an accurately weighed sample of salt in a small test tube. The precipitate can be separated by centrifugation and decantation of the supernatant liquor. The test tube containing the precipitate may be dried in an oven at 120°C. Weigh the precipitate and tube. The procedure of Experiment 5C also can be used. However, the solution of the complex cannot be heated to boiling to coagulate AgCl, because the chloride ligands on cobalt will be displaced.

Part 4: *trans*-Dichlorobisethylenediaminecobalt(III) chloride

Dissolve 1.4 g (0.0063 mole) of cobaltous chloride hexahydrate ($CoCl_2 \cdot 6H_2O$) in 10 ml of distilled water in a 200-mm test tube. To the solution, add 5 ml of a 10% ethylenediamine solution (about 0.008 mole) and mix well. Try not to get the ethylenediamine solution on your skin or clothing because it is toxic. If you do get some on you, wash the area well with water immediately. Fit the test tube with glass tubing and a stopper as shown in Figure 15-6 (and described in Part 1) for bubbling air through the solution. Clamp the assembly in a beaker of water on a ring stand so that it can be heated during the aeration. Bring the water to boiling, and draw air through the test tube with an aspirator for one hour. The oxidation proceeds faster at the higher temperature, and since ethylenediamine is not very volatile, the ethylenediamine will not be lost during aeration. Ammonia, as in Parts 1 and 2, would be lost readily. Because evaporation of the reaction solution will occur during the aeration, add distilled water from time to time to keep the volume of the solution between 3 and 5 ml.

After one hour of aeration, remove the test tube from the water bath, and immediately add 5 ml of concentrated hydrochloric acid. The dark red solution should turn dark green. Cool the test tube in ice water. If the dark red solution does not turn green as the hydrochloric acid is added, concentrate the solution in an evaporating dish on a beaker of boiling water (as in Part 2), and add 1 ml of concentrated hydrochloric acid before cooling. The dark green crystals are the hydrogen chloride salt of *trans*-dichlorobisethylenediaminecobalt(III) chloride. In the following equation, "en" represents ethylenediamine

$$2CoCl_2(aq) + 4H_2N-CH_2-CH_2-NH_2 + 4HCl + \frac{1}{2}O_2$$
$$\rightarrow 2[Co(en)_2Cl_2]Cl \cdot HCl + H_2O$$

Filter the crystals on a sintered glass filter crucible. (Paper can be used if the hydrochloric acid is not too concentrated.) If the yield is small (less than half a gram), it may be advisable to concentrate the solution from which the crystals were obtained. Another crop probably will result.

Place the green plates of *trans*-$[Co(en)_2Cl_2]Cl \cdot HCl$ in a 150-mm test tube with 5 ml of methyl alcohol, crush the crystals, and stir to form a slurry. Heat the test tube in a beaker of water, gently at first, until the methyl alcohol is gone. Then heat the test tube for 15 minutes in boiling water. No more hydrogen chloride should be given off. This can be tested by holding a piece of moistened blue litmus paper over the mouth of the test tube. The product can be analyzed as in Part 3. Weigh the product, and calculate the maximum yield theoretically possible from 1.5 g of cobaltous chloride, as well as the percentage your yield is of the theoretical yield.

Part 5: *cis*-Dichlorobisethylenediaminecobalt(III) chloride

Place 0.5 g of the *trans*-$[Co(en)_2Cl_2]Cl$ in a small evaporating dish, and add 5 ml of water. Heat the dish on a beaker of boiling water so that the solution evaporates to dryness. Scrape the resulting blue-green solid onto a small sintered glass filter crucible with the suction off. (A paper filter on a Hirsch funnel can be used, but not as well.) Add three drops of iced distilled water to the solid, wetting all of the solid. Stir with a stirring rod if necessary. After 20 seconds, turn on the suction. The filtrate contains the green trans compound, and the violet cis compound is left on the filter. The yield of the cis isomer is likely to be only 0.1-0.2 g, but the green solution can be evaporated again producing more of the violet cis compound. This product can be analyzed as in Part 3.

Part 6: Isomerization of *cis*-$[Co(en)_2Cl_2]Cl$ to *trans*-$[Co(en)_2Cl_2]Cl$

Dissolve a small amount of the violet cis compound in a small test tube (100-mm), half full of methyl alcohol. Warm the test tube in a beaker of hot water. Note the color change. Explain.

If a colorimeter is available, the reaction rate can be followed quantitatively. Also, the absorption spectrum of the compounds can be measured by plotting the absorbances measured at each wavelength setting of the colorimeter. The rate is measured best at 540 mμ (millimicrons). About 0.015 g of cis compound in 5 ml of methyl alcohol gives an absorbance of 0.8-1.0 at this wavelength. The rate of the reaction depends on the temperature, the amount of water in the methyl alcohol, and the purity of the compound. The inside of the spectrophotometer cell compartment is warmer than the room, about 35°C. Warm 5 ml of methyl alcohol to this temperature, and dissolve about 0.015 g of the powdered cis compound. Cover the top of the test tube (aluminum foil is satisfactory) to reduce evaporation. Record the absorbance, A, every 10 minutes. After one half-life, that is, when the absorbance has decreased by half, warm the tube in water at about 50°C (methyl alcohol boils at 65°C) for the same period of time that one half-life required at 35°C. Add more methyl alcohol if the level of methyl alcohol has decreased by evaporation. Measure and record the absorbance. This absorbance represents the absorbance at time infinity, A_∞. Plot the logarithm of $(A - A_\infty)$ *versus* time. The slope is $-k/2.3$. The half-life is equal to $t_{1/2} = 0.693/k$.

SUPPLEMENTARY READING

Dickerson, Gray, and Haight (Chapter 10)

Masterton and Slowinski (Chapter 19)

Sienko and Plane (Chapter 12, Section 6; Chapters 19 and 21)

Mahan, *College Chemistry* (Chapter 15, Sections 11 and 12); *University Chemistry* (Chapter 16, Sections 11 and 12)

Brescia, *et al.* (Chapter 25)

Brown (Chapter 20)

Pauling (Chapters 23 and 24)

See also F. Basolo and R. Johnson, *Coordination Chemistry* (W. A. Benjamin, New York, 1964).

QUESTIONS

1. In order to think about complex ions and to consider their reactions, it is necessary to visualize their structures, and this is facilitated by a little experience in drawing "pictures" of them. Therefore, draw equations for all the transformations in this experiment, showing the octahedral structures by the convention illustrated in Figures 15-1 and 15-5.

2. In the hexaamminecobalt(III) chloride, are all the ammonia ligands identical? If not, discuss.

3. In the purpureocobaltic chloride, how many kinds of chloride are there? How do they differ in reactivity? How can this be explained by the structure of the complex ion?

4. How will the solubility of hexaamminecobalt(III) chloride be affected by addition of hydrochloric acid to a solution, as done in the experiment?

PROBLEMS

1. The absorption spectrum of the hexaamminecobalt(III) cation is shown in Figure 15-8. Corresponding absorption maxima of the blue hexaaquocobalt(III) cation are at 6020 and 4020 Å, and those of the yellow hexacyanocobaltate(III) anion are at 3110 and 2590 Å. Calculate the energy (in wave numbers, cm^{-1}) of the corresponding d-electron transitions in these complexes. Discuss the differences in terms of crystal field or ligand field theory. Compare the field strength of the three ligands. Also compare the energy of the d-d absorption in the chloropentaamminecobalt(III) chloride (Figure 15-8) with that of aquopentaamminecobalt(III) cation, which has a maximum at 4880 Å. From this relationship, place chloride in the series of field strengths determined above.

2. Judging from the color of the carbonatotetraamminecobalt(III) chloride, estimate the wavelength of the light absorbed by this com-

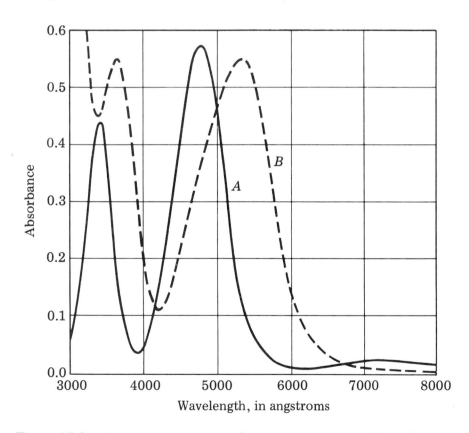

Figure 15-8. Absorption spectra of hexaamminecobalt(III) chloride (Curve A, 0.010M in water) and chloropentaamminecobalt(III) chloride (Curve B, 0.011M in water) from 3000 to 8000 Å, taken with a Cary 14 recording spectrophotometer.

pound. If a spectrophotometer (either manual or recording) is available, measure the absorption spectrum of this compound and that of the diaquotetraamminecobalt cation formed from it.

3. Draw structural representations for all the transformations of the octahedral cobalt species that occur in these syntheses.

4. Make models of the hexamminecobalt, chloropentaamminecobalt, and *cis-* and *trans*-dichlorotetraamminecobalt(III) cations, as well as the square coplanar tetraamminecopper(II) cation and *cis-* and *trans*-dichlorodiammineplatinum(II) molecules. Models can be made from styrofoam balls, applicator sticks or pipecleaners, and glue. Push sticks into the styrofoam ball at the desired angles (and glue, if a permanent model is desired). Molecular models such as the Benjamin/Maruzen HGS Molecular Models (W. A. Benjamin, New York) are inexpensive and very easy to use because hexacoordinate atoms are supplied. Also construct three different octahedral complexes of the formula M(en)$_3$, where en is a chelating ethylenediamine ligand. Pipe cleaners with both ends stuck into a styrofoam ball at octahedral angles can be used to represent the chelating groups. With the Benjamin/Maruzen models, pipe cleaners can be used to connect plastic tubes, which then represent the nitrogen ends of the ethylenediamine ligands.

EXPERIMENT 16

Equipment and supplies

One 50-ml beaker

Burner, wire gauze, ring stand

One funnel

One Hirsch funnel

One filter flask and fitting

One spatula

Analytical balance

One 50-ml burette

Two 250-ml Erlenmeyer flasks

0.5 g Aluminum alloy

15 ml 20% Potassium hydroxide

Glass wool

9 g Oxalic acid

35 ml Ethanol

Filter paper, 11 or 15 cm

Filter paper for Hirsch funnel

1.6 g Potassium permanganate

10 ml $6N$ ($3M$) Sulfuric acid

Time requirement

Four hours: one and one-half to two laboratory periods

POTASSIUM TRIOXALATOALUMINATE TRIHYDRATE

In this experiment, potassium trioxalatoaluminate trihydrate is prepared, and purified by filtration and crystallization.

INTRODUCTION

A sample of an aluminum alloy probably containing some copper is dissolved in potassium hydroxide, forming aluminate ion [$Al(OH)_4^-$] and hydrogen. Oxalic acid is added to form the trioxalatoaluminate. Then the hot solution is filtered to remove insoluble material such as copper metal, or possibly calcium oxalate. The hot solution cools, and the product, potassium trioxalatoaluminate(III) trihydrate, $K_3Al(C_2O_4)_3 \cdot 3H_2O$, precipitates. Although there are exactly three times as many potassium ions as aluminum ions in potassium trioxalatoaluminate trihydrate, this salt crystallizes from a solution which is $0.31M$ in potassium ion and $0.10M$ in trioxalatoaluminate ion. The excess of $0.01M$ potassium ion, as well as impurities such as cupric ion or chloride ion, remains in the solution.

The second part of the experiment is the determination of the oxalate content of the material that you have prepared, thus substantiating the formula $K_3Al(C_2O_4)_3 \cdot 3H_2O$.

The structure of the trioxalatoaluminate anion has an octahedral geometry (Figure 16-1).

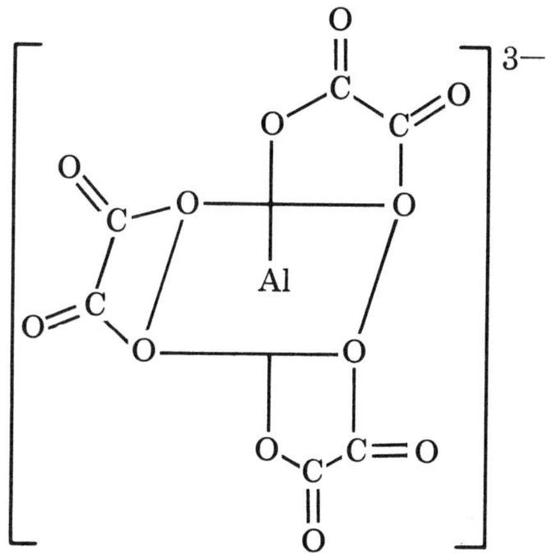

Figure 16-1. Trioxalatoaluminate(III) anion.

Write a balanced equation for the oxidation of aluminum in basic solution. Then calculate the concentration of each of the following constituents present in the solution just prior to the addition of ethanol: K^+, $Al(C_2O_4)_3^{3-}$, $C_2O_4^{2-}$, and OH^-. Assume that the equilibrium constant for the reaction

$$Al^{3+} + 3C_2O_4^{2-} \rightarrow Al(C_2O_4)_3^{3-}$$

is large compared to unity.

Write the balanced equation for the oxidation of oxalate ion by permanganate ion in hot acid solution. Calculate the volume of $6M$ H_2SO_4 required to make 100 ml of $0.2M$ H_2SO_4.

PROCEDURE

Part 1: Preparation of potassium trioxalatoaluminate trihydrate

Weigh about 0.5 g of aluminum alloy into a 50-ml beaker. Cover the metal with 10 ml of warm water and add 15 ml of 20% potassium hydroxide, in small portions, to avoid too vigorous reaction. Heat to boiling to dissolve all of the aluminum, and filter while still hot through a plug of glass wool inserted in the base of a funnel. The residue is mainly copper, and it does not matter that some finely divided material passes into the filtrate at this time, as it will be removed later. Add 5 ml of water to the filtrate and again heat to boiling. Weigh 7 g of oxalic acid (or 10 g of oxalic acid dihydrate), and add this in portions to the hot solution until the precipitate of aluminum hydroxide, which is formed at first, just redissolves. Avoid an excess of acid. Filter the hot solution with suction through paper and, after cooling the filtrate to room temperature, add 25 ml of ethanol. Continue cooling (in ice if necessary) until the complex oxalate commences to crystallize. Upon completion of crystallization, filter with suction and press the cake on the filter firmly with a cork or spatula. After releasing the suction, wash the crystals on the filter with about 10 ml of 50% ethanol-water mixture and return the suction. Wash again with pure ethanol. Dry the crystals, while you proceed with the next section, by gently drawing air through the filter. It would be well to place a clean cloth over the top of the filter to keep off dirt. Weigh the product to the nearest 0.1 g, and calculate a percentage yield on the basis of the formula $K_3Al(C_2O_4)_3 \cdot 3H_2O$.

Part 2: Standardization of 0.02M potassium permanganate

Accurately weigh 0.15-0.20 g oxalic acid, $H_2C_2O_4 \cdot 2H_2O$, and dissolve in 100 ml of 2M H_2SO_4. Heat the solution almost to boiling and titrate with an approximately 0.02M $KMnO_4$ solution, prepared as in Experiment 14. The end point is indicated by a faint pink color that persists for at least one minute. See instructions in Experiment 14 concerning permanganate titrations.

Part 3: Analysis of oxalate

Dissolve an accurately weighed sample (0.2-0.3 g) of the potassium trioxalatoaluminate trihydrate in about 100 ml of 2M H_2SO_4, heat the solution almost to boiling, and titrate with standardized 0.02M potassium permanganate. Calculate the percentage of $C_2O_4^{2-}$ in the complex salt, and compare it to the theoretical value.

SUPPLEMENTARY READING

Dickerson, Gray, and Haight (Chapter 10)

PROBLEM

Oxalate salts sometimes are used as rust removers. The oxalate ion complexes with the iron in the iron oxide, which may be described as $Fe_2O_3 \cdot xH_2O$. These iron-oxalate complexes are soluble. Suggest possible structures for such soluble complexes.

EXPERIMENT 17

Equipment and supplies

One evaporating dish, 00A
One 250-ml beaker
One 50-ml beaker
Burner, wire gauze, ring stand
One 50-ml round bottom flask
One condenser
Rubber or plastic tubing
One Hirsch funnel
One filter flask
One spatula

5 g Sodium cyanate
6 g Ammonium sulfate
15 ml Absolute ethanol
Filter paper
Filter paper for Hirsch funnel
Melting point capillary
Mineral oil for melting point bath

Time requirement

Three hours

PREPARATION OF UREA

INTRODUCTION

Urea was found by Rouelle, in 1773, to be a constituent of urine. It occurs to the extent of 1.5% to 3% in human urine. In 24 hours, 0.51 g of urea is produced per kilogram of body weight. The composition of urea was determined by Prout,[33] in 1818.

In 1828, Freidrich Wöhler prepared urea by heating ammonium cyanate.[34] This was the first organic compound to be prepared from inorganic materials. At this time it was widely believed, even by great scientists, that organic materials could be produced only by living organisms through the action of the "vital force." Wöhler's experiment showed that urea could be formed from inorganic materials without the vital force of a living organism. The vital force theory finally died out about 1850 in chemistry, but continued for a long time in biology.

The synthetic method used in this assignment was used first by Justus von Liebig.[35] He evaporated potassium cyanate and ammonium sulfate to dryness.

PROCEDURE

Dissolve 5 g of sodium cyanate in 15 ml of distilled water in an evaporating dish. Add a solution of 6 g of ammonium sulfate in 20 ml of water to the evaporating dish and stir. Evaporate the solution completely to dryness on a water bath. If a steam bath is not available, set the evaporating dish on a beaker of boiling water. The evaporation takes about 1.5 to 2 hours. Stir occasionally. The reaction is

$$2NaOCN + (NH_4)_2SO_4 \rightarrow 2H_2N\!-\!\underset{\underset{O}{\|}}{C}\!-\!NH_2 + Na_2SO_4$$

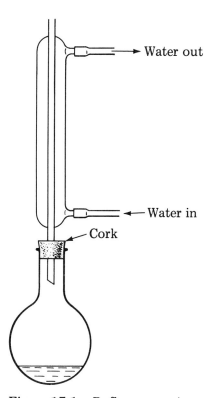

Figure 17-1. Reflux apparatus.

Scrape the solid into a 50-ml flask equipped with a condenser arranged for reflux (see Figure 17-1). Add 15 ml of absolute ethyl or methyl alcohol to the flask by pouring it through the condenser. Boil the mixture for 5 to 10 minutes. Remember that alcohol is flammable. After turning off the flame (if you didn't use a steam bath), filter the hot solution through a fluted filter paper. The extraction can be repeated to get a little larger yield. Cool the beaker containing the filtrate in ice water. Crystals should form. If no crystals form, it may be necessary to evaporate some of the solvent on a steam bath (not with a flame) in the hood. Filter with suction, and let the air suck through the crystals on the filter paper until they are dry. Weigh the product and calculate the percent yield. As a test of the identity of your product, measure the melting point. Urea melts at 132°C. If a melting point block is not available, heat a small amount of the crystals in a small capillary in a beaker of mineral oil. Test the solubility of urea in water. Compare the solubility in water to that in alcohol. Burn some of the urea and some of the starting materials on the tip of a spatula. Do they burn? Is there a residue? Discuss.

SUPPLEMENTARY READING

Dickerson, Gray, and Haight (Chapter 11, Section 8)

Hammond, Osteryoung, Crawford, and Gray (Chapter 11, Section 5, Chapter 14)

EXPERIMENT 18

Equipment and supplies

One porcelain evaporating dish
Burner, wire gauze, ring stand
One conical funnel
One Hirsch funnel
Two 250-ml Erlenmeyer flasks
One filter flask
One 50-ml burette
Analytical balance

10 g Anhydrous sodium hydrogen phosphate

10 ml $1M$ Silver nitrate

Filter paper for Hirsch funnel

35 ml Ethanol

10 ml Acetone

20 ml Glacial acetic acid

10 ml $6M$ Sodium hydroxide

2 ml Phenolphthalein indicator

2 g Oxalic acid

10 g Potassium thiocyanate

2 ml Saturated ferric ammonium sulfate

2 g Silver nitrate

1 g Sodium acetate

20 ml $0.5M$ Silver nitrate

20 ml Aqueous 5% sodium chloride

20 ml $6M$ Nitric acid

Time requirement

Four hours, or two laboratory periods. Other work can be done during the heating period.

DISODIUM HYDROGEN PYROPHOSPHATE

INTRODUCTION

In 1826, Thomas Clark[36] reported the first preparation of sodium pyrophosphate and related compounds in the *Edinburg Journal of Science*. The pyrophosphate was prepared by heating the hydrogen phosphate above 240°C.

The preparation of disodium hydrogen pyrophosphate is accomplished in two steps: thermal conversion of the anhydrous disodium hydrogen phosphate to anhydrous pyrophosphate

$$2Na_2HPO_4 \xrightarrow{heat} Na_4P_2O_7 + H_2O$$

followed by selective acidification in acetic acid to obtain the disodium salt

$$Na_4P_2O_7 + 2H^+ \rightarrow Na_2H_2P_2O_7 + 2Na^+$$

The orthophosphate ion (shown in Figure 18-1 as a hydrogen phosphate) has four oxygen atoms arranged tetrahedrally around a phosphorus atom. The oxygens are in the −2 oxidation state, and phosphorus is in the +5 oxidation state. In the pictorial convention used, the

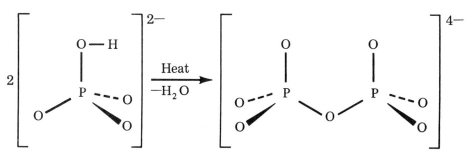

Figure 18-1. Formation of pyrophosphate from hydrogen phosphate.

bond lines that are solid are in the plane of the paper. The bonds directed toward the reader are wedge-shaped, and those angling away from the reader are dashed lines. In the reaction, water is lost, but the phosphorus retains the tetrahedral geometry, one oxygen forming a link between two phosphorus atoms. This ion, $P_2O_7^{4-}$, is called the pyrophosphate ion. Higher polymers exist also, such as tetrametaphosphate, which has a cyclic form.

The analysis, which determines both the pyrophosphate and the replaceable hydrogen, depends on the relative insolubility of the normal silver salt ($Ag_4P_2O_7$) as compared to the acid silver salts such as $Ag_2H_2P_2O_7$. Thus, in the presence of acetate ion (from sodium acetate), silver ion precipitates pyrophosphate quantitatively

$$2OAc^- + H_2P_2O_7^{2-} + 4Ag^+ \rightarrow Ag_4P_2O_7 + 2HOAc$$

After filtration of the silver pyrophosphate, the acetic acid is titrated with standard base. The silver pyrophosphate then is dissolved in nitric acid and the silver content determined by titration with standard thiocyanate.

PROCEDURE

Part 1: Preparation of sodium pyrophosphate decahydrate

Place 10 g of anhydrous disodium hydrogen phosphate in a porcelain dish. Place the dish on a wire gauze and heat it strongly with a Bunsen burner (above 300°C). Continue to heat, mixing the material occasionally, until the addition of a few drops of a dilute solution of silver nitrate to a small sample of the salt dissolved in distilled water yields a pure white precipitate (Ag_3PO_4 is yellow). If the precipitate is yellow, heat the reaction mixture longer. This should require from one to two hours. During this time, start the preparation and standardization of the NaOH and KSCN solutions (Parts 3 and 4).

The cooled anhydrous pyrophosphate is recrystallized by dissolving it in a minimum amount of water (about 30 ml) at 80°C, the temperature of maximum solubility. The hot solution should be filtered with moderate suction. After cooling to ice temperature, filter the crystals and press well. Wash with 10 ml of ethanol and draw air over the product to dry it. Weigh the product, which should be $Na_4P_2O_7 \cdot 10 H_2O$, and calculate the recovery percentage. This makes a convenient stopping point.

Part 2: Preparation of disodium hydrogen pyrophosphate

Dissolve 10 g of the sodium pyrophosphate decahydrate in 20 ml of water at 80°C. Warm 20 ml of glacial acetic acid to 80°C also, mix the two solutions, and maintain this temperature until no further precipitation is apparent. Add 25 ml of ethanol, filter the hot solution with suction, and wash the solid with 10-ml portions of ethanol and then acetone. Release the suction momentarily before pouring the wash solvent over the product. Dry in air and weigh. Calculate a yield of anhydrous disodium hydrogen pyrophosphate ($Na_2H_2P_2O_7$) on the basis of the original Na_2HPO_4.

Part 3: Standardization of 0.1N NaOH

Since this solution must be prepared during the first period and used during the second, it must be stored in a tightly capped bottle. Dilute 6N NaOH to make sufficient 0.1N solution to fill the bottle. Calculate the weight of oxalic acid ($H_2C_2O_4 \cdot 2H_2O$) required to neutralize about 40 ml of 0.1N NaOH. Weigh this amount of acid into a 250-ml flask (in duplicate). The most convenient procedure is to place the acid in a small vial that can be weighed before and after tapping the desired amount of acid into the flask. Thus only three weighings are necessary for duplicate samples. Add 50 ml of water and titrate with 0.1N NaOH, using phenolphthalein as the indicator.

Part 4: Standardization of 0.2N KSCN

Weigh roughly enough potassium thiocyanate (KSCN) to make sufficient 0.2N solution to fill a capped bottle. Standardize against weighed portions of silver nitrate dissolved in 50 ml of water to which 1 ml of ferric ammonium sulfate indicator has been added. The reaction is the formation of insoluble silver thiocyanate. The end point of the titration is indicated by a faint pink color due to ferric thiocyanate ion ($FeSCN^{2+}$), which persists in the liquid even after vigorous shaking.

Part 5: Analysis of the disodium hydrogen pyrophosphate

Dissolve a weighed portion of the anhydrous disodium hydrogen pyrophosphate (0.3-0.4 g) in 50 ml of water, and add about 1 g of sodium acetate. To the well-stirred solution, slowly add a slight excess of 0.5N silver nitrate solution (14-18 ml). Continue to stir until the white precipitate of silver pyrophosphate coagulates. At this point it would be well to add a few more drops of silver nitrate solution to ensure that the precipitation is complete. Filter and wash the beaker and precipitate with several portions of distilled water to collect all of the acetic acid solution with the filtrate. Furthermore, the excess silver nitrate must be washed out of the filter cake as thoroughly as possible. It is not necessary to effect a complete transfer of the silver salt to the filter. Reserve all of the silver salt for subsequent analysis.

To the filtrate, add 5% aqueous sodium chloride until precipitation of excess silver ion is complete. Titrate with standard 0.1N NaOH, using phenolphthalein.

Dissolve all of the silver pyrophosphate with three or four 10-ml portions of hot 3N nitric acid, followed by 10-20 ml of water; collect all of the washings in the filter flask. Add 1 ml of ferric ammonium sulfate indicator, and titrate with potassium thiocyanate. Run a blank, as in Experiment 25. This titration is discussed in detail in that experiment.

Calculate the weight percent of replaceable hydrogen and pyrophosphate, and compare it to the theoretical values.

The acid dissociation constants for pyrophosphoric acid ($H_4P_2O_7$) are K_1, 1.4×10^{-1}; K_2, 1.1×10^{-2}; K_3, 2.1×10^{-7}; K_4, 4.0×10^{-10}. Do these values suggest that it should be the disodium salt that should form in an acetic acid solution ($K = 2 \times 10^{-5}$ for acetic acid)?

PROBLEMS

1. Draw Lewis (electron dot) structures for the orthophosphate and pyrophosphate ions. Indicate the formal charge on the P and O atoms.

2. What orbitals must be used in bonding by P to give the tetrahedral geometry?

3. White phosphorus, P_4, can be oxidized in excess oxygen to the molecular oxide, P_4O_{10}. This oxide is the anhydride of phosphoric acid. In phosphoric anhydride, each phosphorus atom is bonded to four oxygen atoms. Draw a Lewis structure for P_4O_{10} and also indicate the formal charges on the various P and O atoms. Draw a structure showing the geometry of the P_4O_{10} molecule, using a dash to represent a bond. If model kits are available, make a model.

4. Write a mechanism showing how water in excess may react with P_4O_{10}, giving metaphosphates, pyrophosphates, and orthophosphates.

5. In the orthophosphate ion, each P—O bond length is 1.55 Å. In P_4O_{10}, the P—O bond lengths are 1.65 Å, for the oxygen atoms bonded to two P atoms, and 1.39 Å for the oxygen atoms attached to only one P atom. Does this indicate some double bond character in some of the P—O bonds? What orbitals and electrons might be used in such bonding?

6. Give the structure of pyrosulfuric acid. Discuss the geometry of the S atoms.

EXPERIMENT 19

Equipment and supplies

Clock with second hand

One 250-ml beaker

One 400-ml beaker

One 600-ml beaker

One square of cardboard

One special 0.1°C thermometer, −1°C-50°C

Rubber tubing section (see procedure)

Burner, wire gauze, ring stand

16 ml 6M Hydrochloric acid

16 ml 6M Nitric acid

16 ml 6N (3M) Sulfuric acid

16 ml 6M Acetic acid

64 ml 6M Sodium hydroxide

0.5 ml Concentrated sulfuric acid

1 g Anhydrous sodium sulfate

1 g Anhydrous potassium sulfate

Time requirement

Two hours

HEAT OF REACTION AND HEAT OF SOLUTION

The purpose of this experiment is to determine the heats of reaction of strong acids and weak acids with a strong base, and also to investigate the heat of solution of some salts.

PROCEDURE

Part 1: Heat capacity of the calorimeter

Construct a calorimeter as in Experiment 6 and proceed as in Part 1 of the procedure. Calculate C_{cal}, the heat capacity of the calorimeter. The section on heat capacity of the calorimeter, in Part 1 of the procedure in Experiment 6, calibrates the calorimeter for change of temperature from room temperature toward 0°C. For this experiment, it would be better to add 100 g of water at 50°C to the 100 g of water at 25°C in the calorimeter, thus calibrating the calorimeter for the temperature range from room temperature to 50°C.

Remember to handle the 0.1°C thermometers gently! Take the precaution suggested in Experiment 6.

Part 2: Heat of neutralization

Place 100 ml of 1.0M hydrochloric acid in the calorimeter. The temperature should be near room temperature. Prepare 100 ml of 1.0M sodium hydroxide freshly (so that there is little sodium carbonate in it). Record the temperature of the sodium hydroxide, which should be as close to that of the hydrochloric acid as possible. Commence swirling the calorimeter, reading the thermometer every 30 seconds for 20 minutes. Then add the 100 ml of 1.0M sodium hydroxide. Continue swirling, and read the temperature every 30 seconds for 5 minutes. Make a graph of the data and extrapolate, as before, back to the time of mixing. Record the temperature rise.

Repeat this experiment using 100 ml of 0.5M (1.0N) sulfuric acid or 100 ml of 1.0M nitric acid with the 1.0M sodium hydroxide. Proceed in the same way, using 100 ml of 1M acetic acid.

Calculation

The heat of reaction, ΔH, can be obtained from the temperature rise by the following equation, which is written as though the starting temperature for acid and base were 25°C.

$$-\Delta H = mC(T_3 - 25°C) + C_{cal}(T_3 - 25°C)$$

The density of $1M$ hydrochloric acid is only slightly greater than that of water. A 5% solution (about $1M$) has a density of 1.023 g ml^{-1}. Since this will be less than our other errors (*probably*), we shall simplify our calculations and assume that the densities of the $1M$ solutions are the same as that of water. The specific heats of dilute solutions of this concentration are only about 5% lower than the specific heat of water. For example, $1M$ sodium hydroxide has a specific heat of 0.949 cal g^{-1}. Consequently, the errors then involved in the values used for m and C tend to compensate. With these assumptions, calculate the heat of reaction in calories per mole of hydronium ion

$$H_3O^+ + OH^- \to 2H_2O + \text{heat}$$

The heat of reaction per mole, ΔH, is defined as negative in an exothermic reaction. Compare the results of the experiments. Explain similar heats of reaction or differences in heat of reaction.

PROBLEMS

1. Considering these data, how would you expect a rise in temperature to affect the ionization constant of water?

2. If the heat of formation of water (H_f) is -68.3 kcal mole^{-1}

$$H_2(g) + \frac{1}{2}O_2(g) \to H_2O(l) + 68.3 \text{ kcal mole}^{-1}$$

and if the heat of formation of the hydronium ion, $H_3O^+(aq)$, is arbitrarily assumed to have the same heat of formation as water, calculate the heat of formation of the hydroxide ion, $OH^-(aq)$.

Part 3: Heats of solution

In this section, the heat evolved when a compound is dissolved in water will be investigated qualitatively, that is, the sign (+ or −) will be determined. Add about 1 ml of water to a test tube. Then very cautiously, pointing the mouth of the test tube away from yourself and others, add 3 to 5 drops of concentrated sulfuric acid with a medicine dropper. Feel the bottom of the test tube. Does the test tube get warmer, or cooler, or stay about the same temperature?

Put a gram or two of sodium sulfate (Na_2SO_4) into a test tube, and add about 2 ml of water to it. Mix and note any temperature change. Do the same with potassium sulfate (K_2SO_4). Try to explain these phenomena, and the relationships between them.

SUPPLEMENTARY READING

Dickerson, Gray, and Haight (Chapter 4, Section 5, and Chapter 14, Section 3)

Masterton and Slowinski (Chapter 4)

Sienko and Plane (Chapter 14, Section 4)

Mahan (Chapter 8, Sections 3 and 4)

Brescia, *et al.* (Chapter 6)

Brown (Chapter 9, Sections 4 and 5)

Pauling (Chapter 5, Section 7)

Hammond, Osteryoung, Crawford, and Gray (Chapter 9, Section 4)

QUESTIONS

1. Why is heat given off in these reactions of acids and bases? (What happens in these reactions?)
2. Are the heats of neutralization of hydrochloric acid and nitric acid the same? Why is this so?
3. Is the heat of neutralization of acetic acid the same as the other acids? Why might this be so?

PROBLEMS

1. According to a handbook, the heat of solution of sulfuric acid is +17.75 kcal per formula weight, at 18°C. Estimate the temperature rise for the addition of 1 g of sulfuric acid to 1 g (1 ml) of water. DO NOT TRY THIS IN THE LABORATORY! IT IS DANGEROUS! What would you predict would happen?

2. A sample of solid naphthalene ($C_{10}H_8$) weighing 0.600 g is burned to $CO_2(g)$ and $H_2O(l)$ in a constant-volume bomb calorimeter. In this experiment, the observed temperature rise of the calorimeter and its contents is 2.255°C. In a separate experiment, the total heat capacity of the calorimeter and its contents was found to be 2550 cal deg^{-1}. What is ΔE for the combustion of 1.00 mole of naphthalene? What is ΔH per mole?

3. Calculate the standard heat of formation of hydrogen peroxide (H_2O_2) from the heat of reaction for the decomposition

$$H_2O_2(l) \rightarrow H_2O(l) + \frac{1}{2}O_2(g) + 23.48 \text{ kcal}$$

from the standard heat of formation, ΔH_f^0, for liquid water, which is −68.32 kcal mole^{-1}. The heat of the reaction is $\Delta H^0 = -23.48$ kcal mole^{-1}.

4. If you have studied relationships involving free energy and entropy, the following problem may be interesting. Calculate the enthalpy and free energy of the following reaction for Cu_2O, Fe_2O_3, and Al_2O_3 at 25°C.

$$MO_n + nH_2 \rightarrow M + nH_2O$$

The enthalpies and free energies of formation, respectively, for the metal oxides at 25°C are Cu_2O, −39.84, −34.98; Fe_2O_3, −196.5, −177.1; and Al_2O_3, −399.09, −376.77 kcal mole^{-1}. The enthalpy and free energy of formation for water as a gas are −58.74 and −55.83 kcal mole^{-1}, and as a liquid, −69.39 and −57.93 kcal mole^{-1}, respectively. Which reactions will occur at 25°C and at 1000°C? In this problem it will be necessary to assume that the enthalpy and entropy changes are constant over all temperature ranges (ΔH and ΔS are constant).

EXPERIMENT 20

Equipment and supplies

One 200-mm test tube

One special 0.1°C thermometer, −1°C-50°C

One stiff wire for stirrer (about 8 inches)

One two-jaw clamp and stand

Clock with second hand

One graduated cylinder

Analytical balance

One 50-ml Erlenmeyer flask and stopper (for liquid sample)

50 ml Benzene

Ice

1 g Unknown sample (or known)

Time requirement

Two hours

MOLECULAR WEIGHTS BY FREEZING POINT DEPRESSION

The purpose of this experiment is to determine the molecular weight of a substance by the freezing point depression method.

HISTORICAL NOTE

Although the work of J. Blagden (1788) was forgotten and had to be rediscovered, he was the first to notice that there was a simple proportionality between the temperature at which salt solutions freeze and their concentrations. In 1861, Rüdorff[37] discovered the same fact. In 1871, de Coppet also found that similar inorganic salts in equal concentration by moles depressed the freezing point of water to the same degree.

In 1882, F. M. Raoult[38] first noted that aqueous solutions of nonionic organic substances show that solutions of the same molar concentration have the same melting point. Other solvents gave the same result. The general relationship may be expressed as

$$\Delta T_f = k_f m$$

where ΔT_f, k_f, and m are, respectively, the freezing point depression (the difference between the freezing point of the pure solvent and the freezing point of the solution), the freezing point depression constant, and the molal concentration of the solute in the solvent. Molal concentration is the number of moles of the solute in 1000 g of the solvent. Therefore, the relationship can be rewritten as

$$\Delta T_f = k_f \frac{1000 n}{w_s} \quad \text{or} \quad \Delta T_f = k_f \frac{1000 w_x}{w_s \, \text{mol wt}_x}$$

where n is the number of moles of solute, w_s the weight of solvent, w_x the weight of solute, and mol wt_x the molecular weight of the solute.

The connection between the freezing point depression of a solvent and the vapor pressure lowering of solvent by a dissolved solute was deduced theoretically by C. M. Guldberg in 1870. J. H. van't Hoff completed the theoretical derivation (1886) and was able to derive the freezing point constant k_f from thermodynamics in terms of the heat of fusion, H_f, and the freezing point of the solvent T_f

$$k_f = \frac{RT_f^2}{H_f}$$

where R is the gas constant.

PROCEDURE

Benzene will be chosen as a solvent because it has a convenient freezing point, and because many substances are soluble in it. The molal freezing point constant of benzene is 5.12°C. This constant is the depression in freezing point produced by dissolving 1 mole of solute in 1000 g of solvent.

It will be necessary first to find the freezing point of pure benzene. Put about 15 ml of pure benzene into a large dry test tube (200-mm), and suspend the special tenth-degree thermometer so that the bulb is immersed completely in the benzene. (Be very careful with the thermometer; lay it on a paper towel as a cushion.) Adjust a stiff wire for use as a stirrer. Clamp the test tube upright in a beaker of crushed ice and water. The apparatus described is shown in Figure 20-1. Start readings immediately as follows: As the benzene cools, stir it gently but steadily. Read the temperature every 30 seconds. It probably will be necessary to continue the experiment for about 10 minutes. Stir the ice occasionally. Record your data in tabular form. Plot the data, temperature versus time. The freezing point will be indicated by the first region in which the temperature is almost constant for several minutes.

Clean and dry your test tube. With a small graduated cylinder, measure 12 g of benzene. The density of benzene is 0.870 g ml^{-1}. Stopper the benzene to prevent evaporation. Obtain a sample of the unknown substance to be investigated.

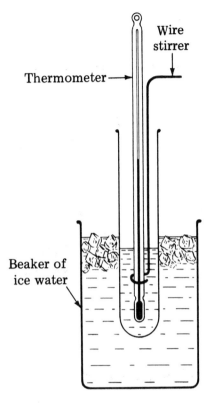

Figure 20-1. Freezing point apparatus.

Solid unknown

If the unknown is a solid, weigh about one gram of the sample on a clean piece of glazed paper. Weigh accurately a 25- or 50-ml Erlenmeyer flask with a stopper. Add the unknown to the flask and reweigh. Add the volume of benzene and reweigh accurately. The flask must be stoppered when the benzene is weighed because benzene evaporates so readily. Dissolve the solid by swirling the flask. Record the weights. Pour the solution into the *dry* test tube and determine the freezing point as before. Record the data in a table as before, and make a plot from the data.

Liquid unknown

If the unknown is a liquid, weigh a 25- or 50-ml flask and stopper accurately. Add to the flask 1.0-1.3 ml or about 1 g of the unknown liquid with a dropper or pipette, and reweigh accurately. Record the weighings. Add the volume of benzene (12 g) to the sample, mix, and reweigh. Pour the solution into the test tube used for the freezing point determination. Determine the freezing point as before. Record the data in a table, and make a plot from the data.

Dispose of the benzene solutions in the sink in the hood, and flush with water.

The freezing point will be the highest temperature at which a cloudy formation of benzene crystals is observed. The temperature probably will not remain constant as in the case of pure benzene, but continue readings for a minute or two after the solution becomes cloudy.

CALCULATIONS

Record the freezing point of the pure solvent, and of the solution. Record the freezing point depression. From the above data calculate the number of moles of sample per 1000 g of solvent. Report the approximate molecular weight of the sample.

SUPPLEMENTARY READING

Dickerson, Gray, and Haight (Chapter 3, Section 5, and Chapter 14, Section 10)

Masterton and Slowinski (Chapter 11, Section 7)

Sienko and Plane (Chapter 8, Section 3)

Mahan (Chapter 4, Section 4)

Brescia, *et al.* (Chapter 3, Section 14; Chapter 13, Sections 18, 19, and 21)

Brown (Chapter 10, Section 7)

Pauling (Chapter 17, Sections 7 and 8)

QUESTIONS

1. Describe how the temperature changes with time as pure benzene is uniformly cooled in the range from $20°C$ to $0°C$.
2. Describe how the temperature changes with time as a benzene solution is cooled. Why doesn't the temperature remain constant at the freezing point as the benzene solidifies?
3. What effect on your determination would evaporation of some benzene after weighing make?
4. How could you determine the freezing point depression constant of benzene experimentally?

PROBLEMS

1. A compound analyzes to be $C_{12}H_{18}O_4Pd$. The freezing point of a 1.2 weight percent solution in bromoform was found to be $2.07°C$. The

freezing point of pure bromoform used was 2.33°C. The molal freezing point depression constant for bromoform is 14.4°C. Calculate the molecular weight of the compound and give the molecular formula.

The freezing point depression of a 3.0 weight percent solution of the same compound in acetic acid was found to be 0.43°C. The acetic acid freezes at 17.0°C and its molal freezing point depression constant is 4.7°C. Calculate the molecular weight and the molecular formula of the compound. Suggest an explanation for the difference.

2. If 1.370 g of sulfur is dissolved in 40.00 ml of carbon disulfide (CS_2), the solution boils at 46.550°C. Pure CS_2 (density 1.263 g ml^{-1}) boils at 46.300°C. The molal boiling point elevation constant for CS_2 is 2.34°C. What is the molecular formula of sulfur?

Suggest an explanation for the atomic weight for sulfur of 93.74 at 230°C obtained by Dumas by the vapor density method (see the historical note in Experiment 7).

EXPERIMENT 21

Equipment and supplies

One 5-ml transfer pipette (Method A)

One 3-ml transfer pipette (Method A)

One 2-ml transfer pipette (Method A)

One 125-ml Erlenmeyer flask and stopper (Method A)

One 5-ml Mohr graduated pipette (Method B)

One pipette bulb

Five 150-mm test tubes and corks

Plastic squeeze bottle

Acetone in squeeze bottle (optional)

One 50-ml burette

Two 125-ml Erlenmeyer flasks

Analytical balance

10 ml Glacial acetic acid

10 ml Ethyl alcohol, absolute

10 ml Ethyl acetate

30 ml $6M$ Hydrochloric acid

150 ml $6M$ Sodium hydroxide

2 g Oxalic acid

1 ml Phenolphthalein indicator

Time requirement

In this experiment, the solutions should be made a week before they are to be used. If the pipette calibrations are done as in Method A, about an hour is needed just for that. The time for the analysis is the same for either method, about two hours. In making the solutions, about half an hour is needed for measuring out the solutions, but the heterogeneous ethyl acetate-water solution takes about an hour of intermittent shaking to prepare.

HOMOGENEOUS EQUILIBRIUM; THE HYDROLYSIS OF ETHYL ACETATE

In this experiment, the equilibrium constant for the hydrolysis of ethyl acetate will be determined.

INTRODUCTION

As mentioned earlier, L. Wilhelmy[39] was first to quantitatively measure the rate of reaction and to integrate the rate law. This was in 1850. In 1862, Marcellin Berthelot and Péan de St. Gilles[40] reported a study of the hydrolysis of esters, including ethyl acetate. In their experiments, no catalyst was used. In 1863, Guldberg and Waage[41] interpreted these data in a theoretical way, stating their "law of chemical equilibrium" or the law of mass action. The hydrolysis of ethyl acetate in dilute hydrochloric acid was investigated by Jones and Lapworth.[42] In this assignment, similar conditions will be employed. The equilibrium will be approached from the side of ethyl acetate and water, and from the side of acetic acid and ethanol, with $6N$ hydrochloric acid as a catalyst

$$CH_3C(=O)-O-CH_2CH_3 + H_2O \rightleftarrows CH_3C(=O)-OH + HO-CH_2CH_3$$

The heat of reaction is very small; consequently, the equilibrium constant is insensitive to temperature. For this reason, it is not necessary to control the temperature of the reaction.

PROCEDURE

Part 1: Preparation of solutions

Clean and dry five screw cap vials or test tubes with corks (ten, if duplicate runs are to be made). Number them 1, 2, 3, 4, and 5.

Method A

Table 1, under Method B, summarizes the instructions below. Clean a 5-ml pipette by rinsing it a few times with distilled water from a squeeze bottle. Using this pipette, add a 5-ml aliquot of distilled water to Tubes 1 and 2. DO NOT USE YOUR MOUTH in this experiment to fill the pipette; use the pipette bulb. Rinse the pipette with 6N HCl. Weigh a small flask accurately. Using the 5-ml pipette, add 5 ml of 6N HCl to the flask (after rinsing the pipette) and reweigh the flask. Now add 5 ml of 6N HCl to each of Tubes 1-5, using this pipette.

Rinse the pipette with water three times and drain. Rinse the pipette with ethyl acetate so that no water remains. (Another method of cleaning the pipette is to rinse it with water three times and then with acetone from a squeeze bottle. To dry the pipette, attach it to the vacuum line or aspirator.) Weigh a 5-ml aliquot of ethyl acetate in a weighed flask fitted with a cork. Ethyl acetate is volatile, so the flask must be closed in order to weigh it. Now add a 5-ml aliquot of ethyl acetate to Tube 3. Close Tube 3 tightly so that ethyl acetate cannot evaporate. Weigh a stoppered flask (or the same one) and add 3 ml of pure acetic acid with a 3-ml pipette (clean) and reweigh. Add 3 ml of acetic acid to Tube 4. Using a *clean* 2-ml pipette, add 2 ml of ethanol to Tube 4 and close it tightly. Determine the weight of 2 ml of ethanol from the 2-ml pipette.

Deliver 2 ml of acetic acid into Tube 5, using the cleaned 2-ml pipette. Weigh a 2-ml aliquot of acetic acid from the 2-ml pipette. Add 3 ml of pure ethanol to Tube 5, using a clean 3-ml pipette. Weigh a 3-ml aliquot of ethanol from the 3-ml pipette. Close the tube tightly.

Two students may collaborate to the extent of each doing half of the calibrations of the pipette for the various liquids. Each student should have his own set of Tubes 1-5, though.

Shake the tubes well. Shake the tubes with the heterogeneous mixtures for a few minutes. Let the tubes stand for a week, if possible, shaking them now and then.

Method B

If the time available is insufficient to do all the weighings described in Method A, less accurate results can be obtained by assuming that the pipettes deliver their nominal volumes. Using densities of the liquids and solutions given in handbooks, and the nominal volumes, the weights of each reactant can be obtained

$$\text{density} \times \text{volume} = \text{weight}$$

A further simplification of the procedure for making the solutions can be made by using a 5-ml Mohr graduated pipette. Of course, this will introduce further error into the experiments. Table 21-1 summarizes the following instructions.

With a clean dry 5-ml pipette, add 5 ml of distilled water to Tubes 1 and 2. After rinsing the pipette with 6M hydrochloric acid, add 5-ml aliquots of 6M hydrochloric acid to Tubes 1-5. After rinsing the pipette with distilled water and then acetone from a squeeze bottle, dry it with a current of air by connecting it to the vacuum line. Then add a 5-ml aliquot of ethyl acetate to Tube 3. Using a clean and dry 5-ml graduated Mohr pipette, add a 3.0-ml aliquot of pure acetic acid to Tube 4 and a 2.0-ml aliquot to Tube 5. With a clean Mohr pipette, add a 2.0-ml aliquot of ethyl alcohol to Tube 4 and a 3.0-ml aliquot to Tube 5. Close all the tubes tightly immediately after adding the samples. Shake the tubes well. Shake the heterogeneous mixtures for a few minutes. Let the tubes stand for a week, if possible, shaking them now and then.

Table 21-1

Tube no.	Water	6M HCl in H_2O	EtOAc	HOAc	EtOH
1	5.0 ml	5.0 ml	0	0	0
2	5.0 ml	5.0 ml	0	0	0
3	0	5.0 ml	5.0 ml	0	0
4	0	5.0 ml	0	3.0 ml	2.0 ml
5	0	5.0 ml	0	2.0 ml	3.0 ml

Part 2: Analysis of method A or B, one week later

Standardize a 2M solution of sodium hydroxide with weighed samples of oxalic acid as in Experiment 13. Determine the concentration of sodium hydroxide to 1% accuracy. About 250 ml of standard solution will be needed for five tubes, so start with 400 ml.

Open a tube (or vial) and pour its contents into a 125-ml Erlenmeyer flask. Rinse the tube with distilled water from a squeeze bottle. Add a few drops of phenolphthalein indicator solution. Now titrate the acid with the standardized sodium hydroxide. *Swirl continuously* as you titrate to the first pink color. The end point will fade in a few seconds, however, because of the reaction of ethyl acetate with sodium hydroxide, which uses up hydroxide ion

$$\text{EtOAc} + \text{OH}^- \rightarrow {}^-\text{OAc} + \text{EtOH}$$

Consequently, titrate quickly, not dropwise. Do not try to get a permanent end point. Treat all of the tubes in this way.

Phenolphthalein is a good indicator for color-blind students. However, it does turn color on the basic side of neutral (pH 8.3-10), and the resulting basic solution does hydrolyze the ester. This problem can be alleviated by employing an indicator that changes color at a somewhat lower pH, such as bromthymol blue (pH 5-7.6).

CALCULATIONS

Calculate the initial number of moles of the reactants ethyl acetate, ethyl alcohol, and acetic acid ($n^{t=0}$) in Tubes 3, 4, and 5 at the beginning (time equals zero), from the weights of the aliquot and the molecular weights. For Method B, the weight of an aliquot is obtained by multiplying known densities by the nominal volume of the aliquot. The density at 20°C is 1.098 g ml^{-1} for 6M hydrochloric acid; 0.901 g ml^{-1} for ethyl acetate; 1.049 g ml^{-1} for acetic acid; and 0.7893 g ml^{-1} for ethyl alcohol. The formulas are ethyl acetate, $CH_3COOCH_2CH_3$; acetic acid, CH_3COOH; and ethyl alcohol, CH_3CH_2OH.

The number of moles of water present initially, $n_{H_2O}^{t=0}$, in Tubes 3, 4, and 5 is calculated from the weight of water present in the 5.0-ml aliquot of $6M$ HCl. The weight of water present in the 5-ml aliquot of $6M$ HCl is equal to the weight of the $6M$ solution less the weight of hydrogen chloride in it

$$n_{H_2O}^{t=0} = \frac{W_{soln} - W_{HCl}}{18 \text{ g mole}^{-1}}$$

The weight of hydrogen chloride is calculated from the number of moles of HCl, which is determined by titrating Tubes 1 and 2 with sodium hydroxide

$$n_{HCl} = V_{NaOH} M_{NaOH}$$

The number of moles of acetic acid present at equilibrium, n_{HOAc}, in Tubes 3, 4, and 5 is equal to the total number of moles of acid in the tube, less the number of moles of HCl in the tube. The total acid is determined by titrating the equilibrated solution in the tube

$$n_{HOAc} = (\text{total moles of acid}) - n_{HCl}$$

In Tube 3, only ethyl acetate and water are present initially with the HCl catalyst. The reaction for attaining equilibrium can be written

$$CH_3\overset{O}{\overset{\|}{C}}-OCH_2CH_3 + H_2O \xrightarrow{H_3O^+} CH_3\overset{O}{\overset{\|}{C}}-OH + HOCH_2CH_3$$

The number of moles of ethyl acetate left at equilibrium (n_{EtOAc}) will be equal to the number at the beginning ($n_{EtOAc}^{t=0}$) less the number of moles of acetic acid found at equilibrium in Tube 3

$$n_{EtOAc} = n_{EtOAc}^{t=0} - n_{HOAc}$$

This is because a mole of acetic acid is formed by the reaction of one mole of ethyl acetate with one mole of water. The number of moles of ethanol must be the same as that of acetic acid because, according to the equation, each mole of ethyl acetate that reacts with one mole of water makes one mole of acetic acid and one mole of ethanol

$$n_{EtOH} = n_{HOAc}$$

The number of moles of water is

$$n_{H_2O} = n_{H_2O}^{t=0} - n_{HOAc}$$

In Tubes 4 and 5, the reverse reaction takes place

$$CH_3\overset{O}{\overset{\|}{C}}-OH + HOCH_2CH_3 \rightarrow CH_3\overset{O}{\overset{\|}{C}}-OCH_2CH_3 + H_2O$$

Consequently, the number of moles of acetic acid originally present, less the number found at equilibrium, is the number of moles of ethyl acetate present at equilibrium. This number is also the number of moles of water formed by the reaction. The number of moles of ethanol is the number of moles of ethanol initially present, less the number of moles of acetic acid that reacted with the ethanol making ethyl acetate.

$$n_{EtOAc} = n_{HOAc}^{t=0} - n_{HOAc}$$

$$n_{H_2O} = n_{H_2O}^{t=0} + n_{EtOAc}$$

$$n_{EtOH} = n_{EtOH}^{t=0} - n_{EtOAc}$$

The equilibrium expression for the equilibrium

$$CH_3\overset{O}{\overset{\|}{C}}-OCH_2CH_3 + H_2O \rightleftarrows CH_3CH_2OH + CH_3\overset{O}{\overset{\|}{C}}-OH$$

is written

$$K = \frac{[CH_3CH_2OH][CH_3COOH]}{[CH_3COOCH_2CH_3][H_2O]}$$

Since, in a particular tube, the volume of the solution is the same for all quantities, K can be written

$$K = \frac{n_{CH_3CH_2OH} \, n_{CH_3COOH}}{n_{CH_3COOCH_2CH_3} \, n_{H_2O}}$$

where n is the number of moles of the components at equilibrium.

Calculate K for each equilibration that you performed. Average the values from equilibrations, where the catalyst concentration was $6M$.

SUPPLEMENTARY READING

Dickerson, Gray, and Haight (Chapter 15)

Masterton and Slowinski (Chapter 12)

Sienko and Plane (Chapter 11)

Mahan (Chapter 5)

Brescia, *et al.* (Chapter 14)

Brown (Chapter 14)

Pauling (Chapter 18)

Hammond, Osteryoung, Crawford, and Gray (Chapter 10)

QUESTIONS

1. What role does the hydrochloric acid play in the equilibrium?
2. What is the initial reaction that occurs in Tubes 4 and 5?
3. What is the initial reaction in Tube 3? Express this with an equation.
4. In the solution at equilibrium (either 3, 4, or 5), what reactions are taking place?
5. How could you determine the rate constant and the energy of activation for the reaction of pure ethyl acetate with water in the presence of water to form acetic acid and ethanol?
6. How could you show that the heat of reaction for this reaction is nearly zero?
7. Suppose that you wanted to make perdeuteroethyl acetate from deuteroacetic acid and deuteroethanol by the method used here

$$D_3CCOOD + D_3CCD_2OD \xrightarrow{D^+} D_3CCOOCD_2CD_3 + D_2O$$

The deuteroethanol is much more expensive than the deuterated acetic acid. How would you make the most ethyl acetate most cheaply?

EXPERIMENT 22

Equipment and supplies

One conductance bridge (see procedure and Reference 47—total cost $18.00)

One 10-ml graduated cylinder

One 100-ml graduated cylinder (5-ml Mohr pipette may be used)

One 100-ml beaker

10 ml 1.0M Hydrochloric acid

10 ml 1.0M Sodium chloride

10 ml 1.0M Sodium hydroxide

10 ml 1.0M Acetic Acid

Time requirement

One and one-half hour

CONDUCTANCE OF ELECTROLYTIC SOLUTIONS

In this experiment, the ability of a solution of an electrolyte to conduct electricity will be measured. The dissociation constant for acetic acid will be determined.

HISTORICAL NOTE

Faraday introduced the names *ion, cation,* and *anion* with the aid of a clergyman friend who was familiar with Greek. These ions were thought to be formed by the action of the electric current on the electrolyte. Rudolph Clausius[43], in 1857, accounted for the conduction of electric current by electrolytes (either molten or in solution) by considering them dissociated to a slight extent into electrically charged ions, even when there was no electric potential applied to the solution. The ions were the carriers of the electric current when electric potential was applied. He attributed the idea partially to Williamson, who had suggested that ions are intermediates in the formation of ethers (1851). F. Kohlrausch found that the electrical conductivity of strong electrolytes was large, and approached a maximum value with dilution. In contrast, the conductivity of weak electrolytes increased greatly with dilution and showed no tendency toward a constant value at high dilution. Kohlrausch[44] also developed a convenient and accurate method of measuring conductivity, using alternating current and a Wheatstone bridge. The apparatus used in this experiment is similar in principle.

Strong electrolytes were known to give larger values for freezing point depression, boiling point elevation, and osmotic pressure than nonelectrolytes. Van't Hoff introduced the "*i* factor" to correct for this difference.[45] J. Thomsen found the heats of neutralization of strong acids (e.g., HCl, H_2SO_4, and HNO_3) by NaOH to be very nearly the same, about 13.8 kcal equiv^{-1}. The heats of neutralization of weak acids were found to be smaller.

In 1882, Svante Arrhenius presented a theory, based on his studies of electrical conductivity of electrolytes, that strong electrolytes, MX, were almost completely dissociated into ions, M^+ and X^-, in aqueous solution. The theory was considered so outlandish that he very nearly was denied the Doctorate of Philosophy degree. He further developed his theory and published a complete discussion in 1887.[46] He showed that the Van't Hoff i factor was practically the same as the number of ions formed by strong electrolytes. For example, NaCl forms two ions, Na⁺ and Cl⁻, and the i factor is nearly two in water. For $BaCl_2$, it is three. Weak electrolytes were said to be only partly ionized, and the fraction ionized, α, was defined.

Prejudice was very strong against Arrhenius' theory, so that it was not generally accepted. However, all opposition effectively disappeared after Sir William Bragg and his son W. Lawrence Bragg, in 1913, showed by x-ray studies that solid sodium chloride was a lattice of sodium and chloride ions.

INTRODUCTION

The apparatus that will be used in this experiment employs a Wheatstone bridge. The apparatus is, in principle, the same as research instruments designed for these purposes. In the low cost apparatus[47] used here, 60-cycle alternating current (house current) and simple potentiometers are used. A photograph of the apparatus with a conductivity cell is shown in Figure 22-1. A schematic diagram of the instrument appears in Figure 22-2. In the photograph, the dial labeled BAL

Figure 22-1. Wheatstone bridge with conductivity cell.

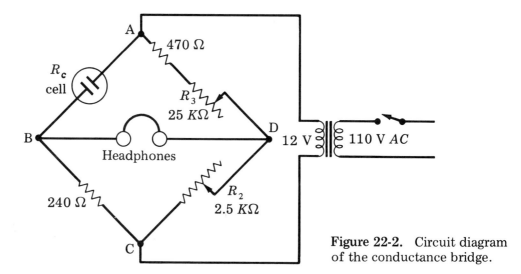

Figure 22-2. Circuit diagram of the conductance bridge.

(balance) corresponds to variable resistance, R_2, and the dial labeled CAL (calibrate) corresponds to variable resistance, R_3. The resistances of R_2 and R_3 vary linearly with the rotation of the dials.

The principles of the operation of the apparatus can be understood using only Ohm's law. Ohm's law states that the voltage (V) between two points on a conductor is equal to the current (I) passed, multiplied by the resistance (R) of the conductor

$$V = IR$$

From Ohm's law, we can derive a relationship for the resistance of the cell, R_c.

When the circuit shown in the diagram (Figure 22-2) is in operation, and when R_3 and R_2, which are variable resistances, are adjusted so that no signal (or a minimum signal) is heard in the headphones, no current is flowing between Points B and D. If no current flows from B to D, the voltage (or potential) at Point B must be equal to the voltage at Point D, which requires that the voltage between Points A and B (V_{AB}) equals the voltage between Points A and D (V_{AD}). Likewise, V_{BC} equals V_{DC}. Although current is not flowing from Point B to Point D, current does flow from A to B to C and from A to D to C. The voltage difference between Points A and B, V_{AB}, is equal to the current, I_{ABC},, multiplied by the resistance of the cell, R_c

$$V_{AB} = I_{ABC} R_c$$

Similarly,

$$V_{AD} = I_{ADC}(470\ \Omega + R_3)$$

where Ω = 1 ohm. Since V_{AB} and V_{AD} are equal, we can write

$$I_{ABC} R_c = I_{ADC}(470\ \Omega + R_3)$$

By the same arguments,

$$V_{BC} = V_{DC}$$

and

$$I_{ABC}\,240\,\Omega = I_{ADC}\,R_2$$

Dividing one equation by the other, we eliminate the current from the relationship

$$\frac{I_{ABC}\,R_c}{I_{ABC}\,240\,\Omega} = \frac{I_{ADC}\,(470\,\Omega + R_3)}{I_{ADC}\,R_2}$$

This can be rearranged to give an equation for the resistance of the cell

$$R_c = \frac{240\,\Omega\,(470\,\Omega + R_3)}{R_2}$$

Resistance, R, is inversely proportional to conductance, Y

$$R_c = \frac{1}{Y_c}$$

Thus,

$$Y_c = \frac{1}{R_c} = \frac{R_2}{240\,\Omega\,(470\,\Omega + R_3)}$$

Since only R_2 and R_3 are variable resistances, and if R_3 is kept constant, then the conductance of the cell is directly proportional to R_2. R_2 is *not* the resistance of the cell but is proportional to the *conductance* of the cell.

In this experiment, the conductance of aqueous acetic acid will be compared to the conductance of the same concentration of aqueous hydrochloric acid. Acetic acid is a weak acid, which means it is only slightly dissociated into free ions

$$HOAc + H_2O \rightleftarrows H_3O^+ + {}^-OAc$$

Reactions of the dissociated ions at the electrodes, and the diffusion of ions in the solution, result in conduction of electric current in the solution. Therefore, the conductance of the solution should be proportional to the concentration of ions. In dilute aqueous solution, hydrochloric acid is dissociated almost completely into free hydronium ions and chloride ions. Conductance of aqueous hydrochloric acid is due to reactions and diffusion of these ions.

If chloride ion conducts as well as acetate ion, we can assume that the conductance of $0.01M$ hydrochloric acid approximates the conductance of $0.01M$ acetic acid, if the acetic acid were 100% dissociated. Hence, the ratio of the actual conductance of $0.01M$ acetic acid to the conductance of 0.01M HCl approximates α, the fraction of acetic acid that is dissociated.

$$\frac{Y^{HOAc}}{Y^{HCl}} = \alpha$$

Substituting the relationship of measured values for Y gives

$$\frac{R_2^{HOAc}}{240\ \Omega\ (470\ \Omega + R_3)} \times \frac{240\ \Omega\ (470\ \Omega + R_3)}{R_2^{HCl}} = \alpha$$

which simplifies to the following, if R_3 is kept constant

$$\frac{R_2^{HOAc}}{R_2^{HCl}} = \alpha$$

The assumption that the conductances of chloride and acetate ions are equal actually is not valid. However, the conductance of hydronium ion is seven or eight times larger than that of chloride or acetate. So this difference between chloride and acetate makes little difference in the overall conductivity, because it is almost all due to hydronium ion.

If the dissociation of acetic acid is represented by the equilibrium

$$HOAc + H_2O \rightleftarrows H_3O^+ + {}^-OAc$$

the following mass action expression can be written

$$K = \frac{[H_3O^+][{}^-OAc]}{[HOAc]}$$

If α is the fraction of acetic acid that dissociates, and C is the formal concentration of acetic acid, then

$$\alpha C = [H_3O^+] = [{}^-OAc]$$

$$(1 - \alpha)C = [HOAc]$$

$$K = \frac{(\alpha C)^2}{(1 - \alpha)C}$$

$$K = \frac{\alpha^2 C}{(1 - \alpha)}$$

Using the equation for α in terms of R_2, the dissociation constant, K, for acetic acid can be calculated from the above equation.

THE CELL

The cell is constructed of two thin platinum squares attached to the ends of glass tubes that are inserted in a rubber stopper. This stopper fits into a 100-ml beaker. Use only 100-ml beakers. The area of each platinum electrode is about 1 cm^2 and the distance between them is about 1 cm. The platinum squares have been spot-welded to platinum wires that are sealed in the ends of soft glass tubes. The electrodes are covered with platinum black, which increases the surface area. The electrodes can be bent easily, and the platinum black rubs off. Consequently, do not touch them with fingers, and be careful not to bump them against the beaker. If the path length (distance between electrodes) changes too much, the bridge may not balance. If the path length is changed between measurements, the experiment must be repeated. Always keep the electrodes wet. They will not work well if they dry out. Keep them immersed in distilled water when not in use.

PROCEDURE

Part 1: The conductance of acetic acid

Before using the apparatus, you should be "checked out" by the instructor. Fill a 100-ml beaker to about one inch from the top with $0.010M$ HCl. Rinse the electrodes by dipping them into a 100-ml beaker of distilled water. Then put the electrodes into the beaker containing the $0.010M$ HCl.

Set variable resistance R_2 at the maximum value. Switch the instrument on (also plug it into the 110 V current outlet if it isn't already). Listening to the sound in the headphones, adjust the variable resistance R_3 until a minimum signal or no signal (null point) is heard. Sixty-cycle hum is not easy to hear, but if you concentrate you will hear a distinct null point at this concentration. Record the setting of R_2. Switch the instrument off.

Fill another 100-ml beaker about an inch from the top with $0.010M$ acetic acid. Rinse the electrodes by dipping them in the 100-ml beaker of distilled water and swirling the beaker and its contents. Put the electrodes into the $0.010M$ acetic acid. Switch the instrument on. Listening to the hum in the headphones, adjust R_2 until a null point is found. Record R_2. DO NOT CHANGE THE CALIBRATION OF R_3. R_2 is a relative measure of the conductance of $0.010M$ acetic acid compared to the conductance of $0.010M$ HCl, as represented by the initial R_2 setting, which was the maximum value of 10.0.

Rinse the electrodes and measure the conductance of $0.010M$ NaCl solution. For this, calculate the ratio of the conductance of NaCl to HCl at this concentration.

Rinse the electrodes and measure $0.001M$ HCl, leaving R_3 the same. Record R_2. The value of R_2 for $0.001M$ HCl should be nearly 1/10 of R_2 for $0.01M$ HCl.

Standardize the bridge against $0.0010M$ HCl as was done for $0.010M$ HCl. Set R_2 at the maximum value of 10.0, and adjust R_3 to the null point. Then rinse and measure $0.0010M$ acetic acid. Record R_2 for the acetic acid solution. Then rinse and measure $0.00050M$ HCl. Record R_2.

Standardize the bridge against $0.00050M$ HCl (or $0.00010M$, if the bridge will balance) and measure $0.00050M$ acetic acid. Record R_2.

SUPPLEMENTARY READING

Dickerson, Gray, and Haight (Chapter 3, Sections 1-5, Chapter 16)

Masterton and Slowinski (Chapter 11, Section 6, Chapter 17)

Brescia, *et al.* (Chapter 15, Sections 14-21)

Brown (Chapter 11, Section 5)

Hammond, Osteryoung, Crawford, and Gray (Chapter 4, Section 2, Chapter 9, Section 5)

CALCULATIONS AND QUESTIONS

1. Calculate α for acetic acid at concentrations $0.01M$, $0.001M$, and $0.005M$. Make a graph of α versus concentration. How does α change with concentration? Rationalize this phenomenon.

2. How does the conductivity of HCl change on dilution? Why?

3. Calculate K, the dissociation constant of acetic acid from these values of α.

4. Calculate the ratio of the conductance of NaCl to that of HCl. Rationalize the difference in conductance between the two.

Part 2: Change in conductance with acid-base neutralization

The change in conductance with progressive neutralization of $0.010M$ HCl with $0.10M$ NaOH will be investigated now.

Place 70 ml of $0.010M$ HCl in a 100-ml beaker. Position the electrodes in the beaker. Switch the conductance bridge on. Adjust R_3 to the null point with R_2 set at the maximum value of 10.0. Switch off the bridge. Now add 1.0 ml of $0.10M$ NaOH, using a pipette (or burette) through the hole above the lip of the beaker. Mix the solution. This may be done by lifting out the electrodes and pouring the solution back and forth into another beaker. A stirring rod can be used. Replace the electrodes in the beaker with the solution, and switch on the bridge. Adjust R_2 to the new null point and record R_2. Switch off the bridge. Repeat this process with ten or eleven 1-ml aliquots of NaOH. Switch the bridge off.

Plot the values of R_2 against the milliliters of $0.10M$ NaOH added. Draw lines through your points and extrapolate to the equivalence point. Remember that R_2 is directly proportional to the conductance of the solution in the cell. Rationalize the change in conductance (or R_2) with addition of NaOH. Note that this phenomenon can be used as a method of detecting the equivalence point in a titration.

Note on platinizing electrodes

Clean the electrodes in hot (steam bath) aqua regia (CAUTION!) and rinse them with distilled water. Keep fingers off the clean metal to avoid getting grease marks on it. Electrolyze the electrodes in 1% chloroplatinic acid with 0.01% lead acetate, using a 1.5 V dry cell. Use a platinum wire as the anode. Space the anode from the electrode so that there is a gentle evolution of gas during the electrolysis. A suitable coating should deposit in 1-2 minutes.

EXPERIMENT 23

Equipment and supplies

Six 150-ml test tubes

One 10-ml graduated cylinder

One 100-ml graduated cylinder

25 ml Unknown solution

16 ml 6M Acetic acid

25 ml 1M Sodium acetate

10 ml 1M Ammonium chloride

1 ml or less of the following indicator solutions (in dropper bottles):

    Methyl violet

    Thymol blue

    Methyl orange

    Methyl red

    Bromthymol blue

    Phenolphthalein

    Thymolphthalein

    Sodium indigodisulfonate

5 ml Each of buffer solutions pH 1-14

Time requirement

Two hours

INDICATORS AND pH

The purpose of this experiment is to determine the pH of various solutions by comparison of indicator colors. These measurements will be used to determine the ionization constants of acetic acid and ammonia.

INTRODUCTION

Indicators are weak acids or bases that change color when converted to the corresponding salts. The salt is a different color from the acid or base form. The exact acidity or basicity at which the color change occurs is a property of the indicator compound.

The equilibrium for a weak acid indicator, HIn, can be represented as

$$HIn + H_2O \rightleftharpoons H_3O^+ + In^-$$

(a color) (different color)

$$K_{In} = \frac{[H_3O^+][In^-]}{[HIn]}$$

When the color is half changed, $[HIn] = [In^-]$. This is considered the end point. In practice, the range of $[H_3O^+]$ over which color changes is usually very narrow compared to the mesurements being made. For phenolphthalein, the range in which there is an observable amount of color change is pH 8-10, that is, $[H_3O^+] = 10^{-8}$ to $10^{-10} M$. For phenolphthalein, the pK_{In} is 8.9, so when $[HIn] = [In^-]$, pH is 8.9. The colors of indicators are ordinarily extremely intense; consequently, only a very small amount is needed. This small amount is not enough to change noticeably the acidity of the solution being measured.

The indicators are compounds of carbon. For example, phenolphthalein has the formula $C_{20}H_{14}O_4$, and the structure shown in Figure 23-1. The intense red color of the basic form is due to the absorption of light in the visible spectrum, whereas the acidic form absorbs in the ultraviolet region of the spectrum. The shift of wave-

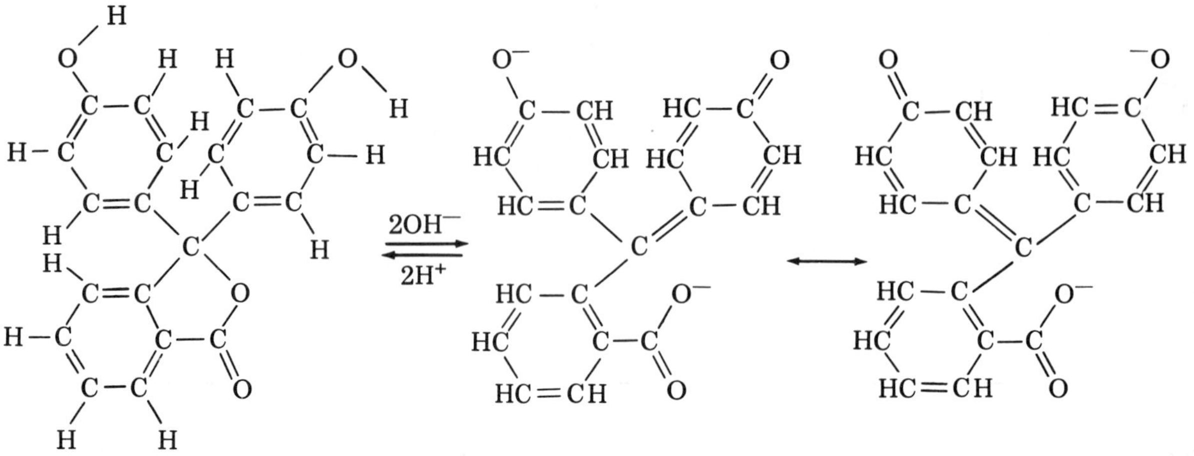

Figure 23-1. Phenolphthalein.

Colorless form ⇌ Red form (2OH⁻ / 2H⁺)

length absorbed, from the ultraviolet to the visible, is the result of increased delocalization of electrons in the molecule after loss of two protons, indicated by the resonance structures.

As you know, water dissociates to a slight extent

$$H_2O + H_2O \rightleftarrows H_3O^+ + OH^-$$

for which the equilibrium expression

$$K_w = [H_3O^+][OH^-] = 10^{-14}$$

can be written. This relationship between $[H_3O^+]$ and $[OH^-]$ is always true for water at 25°C. By definition, the pH is given by the equation

$$pH = -\log[H_3O^+]$$

In this experiment, the dissociation of acetic acid

$$HOAc + H_2O \rightleftarrows H_3O^+ + {}^-OAc$$

will be measured and the dissociation constant evaluated

$$K = \frac{[H_3O^+][{}^-OAc]}{[HOAc]}$$

If the pH is determined, $[H_3O^+]$ can be calculated. If this is a solution of only acetic acid in water, then

$$[H_3O^+] = [{}^-OAc]$$

The acetic acid concentration will be the formal concentration, C, less the concentration of $[H_3O^+]$, which is equal to the amount of acetic acid that dissociated

$$[HOAc] = C - [H_3O^+]$$

The dissociation constant may be calculated from the following equation:

$$K = \frac{[H_3O^+]^2}{C - [H_3O^+]}$$

In many cases, $[H_3O^+]$ will be so much smaller than C that it may be neglected. Thus,

$$C \simeq C - [H_3O^+]$$

PROCEDURE

In clean test tubes, obtain 5 ml portions of each of the side-shelf solutions of pH 1, 2, 3, 4, 5, 6, 7, 8, 9, 10, 11, 12, 13, and 14. The solutions of pH 1 and pH 2 are dilute hydrochloric acid. The solutions of pH 14 and 13 are dilute sodium hydroxide, and can be made from dilute sodium hydroxide. The intermediate pH solutions are buffers. To each solution add one drop of thymol blue indicator. It is important that the volume of solution and the number of drops of indicator be uniform in order to compare colors accurately. Record the colors observed in this experiment, and any others, in a table with 14 columns, for pH 1 through 14. List the indicators vertically. Repeat the procedure using bromthymol blue as the indicator.

The following indicators change color in the pH ranges listed.

1.	Methyl violet	yellow to violet, 0-3
2.	Thymol blue	colorless to yellow, 1-3; yellow to blue, 8-10
3.	Methyl orange	red to yellow, 3-4.5
4.	Methyl red	red to yellow, 4-6
5.	Bromthymol blue	yellow to blue, 5-8
6.	Phenolphthalein	colorless to red, 8-10
7.	Thymolphthalein	colorless to blue, 9-11
8.	Sodium indigodisulfonate	blue to yellow, 12-14

Obtain a sample of an unknown acid or base. Determine its pH as follows. Use bromthymol blue to decide whether the pH is above or below the middle range of the bromthymol blue color change. If it is blue, continue tests with phenolphthalein, thymolphthalein, and sodium indigodisulfonate. If the solution is yellow with bromthymol blue, try indicators that change color at pH's lower than that of bromthymol blue. When the exact range is determined, compare the shade with known buffer solutions and the same indicator. In comparing shades of colors, it is best to look down the two test tubes being compared at a white surface. The heights of liquid in the test tubes should be the same for each indicator. Calculate the hydronium ion concentration in the unknown. Report this [H⁺] and the pH to the instructor.

Using indicators as above, determine the pH of $1.0M$ aqueous acetic acid made by diluting the $6.0M$ reagent. Estimate to 0.5 pH unit. Dilute the $1.0M$ acetic acid to 1:10, making $0.10M$ acetic acid, and determine the pH to 0.5 pH units. Determine the pH of a solution made by mixing 10 ml of $1.0M$ acetic acid and 25 ml of $1.0M$ sodium acetate. Use all of these determinations to compute the dissociation constant of acetic acid.

Determine the pH of $1.0M$ ammonium chloride to 0.5 pH unit. Calculate the hydrolysis constant for the ammonium ion and the constant for ammonia. The reactions involved are

$$NH_4^+ + H_2O \rightleftarrows NH_3 + H_3O^+$$

$$NH_3 + H_2O \rightleftarrows NH_4^+ + OH^-$$

The ionization constant for water, $K_w = 1 \times 10^{-14}$, will be needed also.

SUPPLEMENTARY READING

Dickerson, Gray, and Haight (Chapter 16)

Masterton and Slowinski (Chapters 16, 17, and 18)

Sienko and Plane (Chapter 12)

Mahan (Chapter 5, Sections 2-5)

Brescia, *et al.* (Chapter 19)

Brown (Chapters 12 and 15)

Pauling (Chapter 19)

Hammond, Osteryoung, Crawford, and Gray (Chapter 12, Section 4)

QUESTIONS

1. Why is it important that the volumes and amounts of indicator be identical for each solution being compared?

2. Why don't the indicators, some of which are themselves acids, affect the pH of the solution being measured?

3. In the mass action expression ($K_w = [H_3O^+][OH^-]$) for the dissociation of water to hydronium ions and hydroxide ions, why isn't the concentration of water written?

4. Why doesn't a $1M$ solution of acetic acid have as low a pH (or as high a hydronium ion concentration) as hydrogen chloride in water?

5. Explain why sodium chloride salt (NaCl, the resultant product from the reaction of an equal number of moles of HCl and NaOH) makes an aqueous solution of pH 7.0, while ammonium chloride

salt (NH_4Cl, the result of reaction of equimolar amounts of ammonia, NH_3, and hydrochloric acid, HCl) makes an acidic solution? Both salts are strong electrolytes, being essentially 100% dissociated in dilute water solution.

PROBLEMS

1. Calculate the pH of a 0.10M solution of butyric acid (abbreviated HOBu, but it actually has the structure H_3C—CH_2—CH_2—COOH). The dissociation constant is 1.48×10^{-5}. Also calculate the pOH of the solution.

2. How would you prepare a buffer solution of acetic acid and sodium acetate having a pH of 4.4? Use any amounts of pure acetic acid, solid sodium acetate, and water that you choose. Give detailed instructions with the amounts in weight and volume, not in moles. Acetic acid is $C_2H_4O_2$ and sodium acetate is $NaO_2C_2H_3$.

EXPERIMENT 24

Equipment and supplies

Procedure A

Six 13- × 100-mm test tubes and corks

Centrifuge with head for 100-mm test tubes

One medicine dropper

20 ml 0.1M Copper chloride

10 ml 0.2M Sodium hydroxide

10 ml 1M Ammonia

10 ml 1M Ammonium nitrate

10 ml 6M Hydrochloric acid

Procedure B

Colorimeter (Spectronic 20, Fisher Electrophotometer, Klett, etc.)

One 5-ml Mohr graduated pipette

One pipette bulb

Five 18- × 150-mm test tubes

Four 13- × 100-mm test tubes with corks

One 10-ml graduated cylinder

Centrifuge with head for 100-mm test tubes

10 ml 0.1M Copper sulfate

10 ml 1M Ammonia

10 ml 1M Ammonium nitrate

10 ml 0.1M Copper chloride

12 ml 0.2M Sodium hydroxide

Time requirement

Procedure A: Two and one-half hours

Procedure B: Two hours

THE TETRAAMMINECOPPER(II) CATION

The purpose of this experiment is to measure the formation constant of the tetraamminecopper(II) ion by colorimetry.

INTRODUCTION

Anhydrous copper sulfate ($CuSO_4$) is white, which means that it does not absorb light in the visible region of the spectrum. The hydrated copper sulfate ($CuSO_4 \cdot 5H_2O$) is blue. The structure of the compound can be represented more accurately as $Cu(H_2O)_4 SO_4 \cdot H_2O$ where four water molecules are bound to the copper ion and the fifth is a water of crystallization. The water molecules are arranged at the corners of a square, with the copper at the center as shown in Figure 24-1. Such an arrangement is called square coplanar. The oxygen of each water molecule shares one pair of electrons with the central copper ion. The absorption spectrum of 0.01M copper sulfate is shown, in Figure 24-2, by the dotted line, A. Absorbance is plotted against wavelength in angstroms (Å). Notice that the compound absorbs light of wavelengths from 6000 to above 8000 Å, which is the yellow-to-red region of the visible spectrum. The light transmitted through the solution comes out richer in light of blue wavelengths (4000 to 5000 Å) than white light, and so the solution looks blue.

When ammonia is added to a solution of copper(II) cation, a deep blue color is formed immediately. The blue color is due to the complex ion $Cu(NH_3)_4^{2+}$

$$Cu(H_2O)_4^{2+} + 4NH_3 \rightleftarrows Cu(NH_3)_4^{2+} + 4H_2O$$

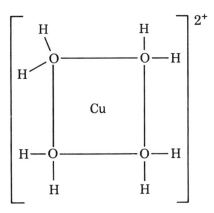

Figure 24-1. Structure of $Cu(H_2O)_4^{2+}$.

This complex ion, the tetraamminecopper(II) cation, has a square coplanar geometry also. The absorption spectrum of this complex ion in 0.05M ammonia is shown in Figure 24-2 as the solid line, B. In this complex, also, the light of yellow and red wavelengths is absorbed more than the blue, so the solution appears blue. The tetraamminecopper(II) cation is the principal copper species present in ammonia solution of concentration 0.01 to 5M. However, at lower concentrations of ammonia, other copper species having 3, 2 or 1 molecules of ammonia

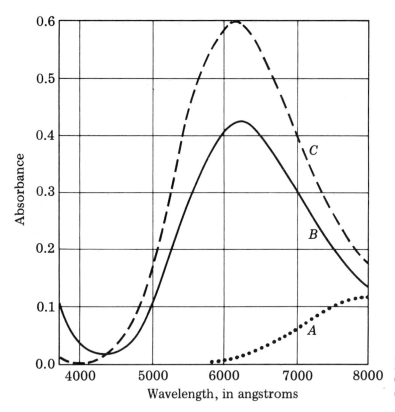

Figure 24-2. Absorption spectrum of $0.01M$ $CuSO_4$ in (A) water; (B) $0.09M$ NH_3 and $0.09M$ NH_4NO_3; and (C) $1M$ NH_3.

may be present. At higher concentrations of ammonia, a pentaamminecopper(II) cation, $Cu(NH_3)_5H_2O^{2+}$, is formed also. In Figure 24-2, the dashed line, C, represents the absorption spectrum of $0.01M$ tetraamminecopper(II) cation in a solution of $1M$ ammonia. It is evident that the absorbance is larger in $1M$ ammonia than in $0.05M$ ammonia, and also, the wavelength for the maximum absorption is slightly different. These differences are due to the formation of perhaps 25% of the pentaamminecopper complex. The tetraamminecopper complex is practically the only copper species present under the conditions of the experiment; an ammonia-ammonium chloride buffer with a $0.05M$ concentration of ammonia.

DERIVATION

Cupric ion reacts with ammonia to form a deep blue complex ion $Cu(NH_3)_4^{2+}$

$$Cu^{2+} + 4NH_3 \rightleftarrows Cu(NH_3)_4^{2+} \qquad (1)$$

The mass action expression for this equilibrium can be written as follows; K_f is the formation constant of the tetraamminecopper(II) ion:

$$K_f = \frac{[Cu(NH_3)_4^{2+}]}{[Cu^{2+}][NH_3]^4}$$

The equilibrium for Reaction 1 lies very far on the side of the complex ion; so far that the cupric ion concentration is too small to measure, except by an electrochemical cell. In this experiment, the formation constant will be measured by studying the equilibrium between solid copper hydroxide and cupric ammine complex in ammonia.

When cupric hydroxide is treated with ammonia, one expects the reaction

$$Cu(OH)_2(s) + 4NH_3 \rightleftarrows Cu(NH_3)_4^{2+} + 2OH^- \qquad (2)$$

The concentration of the cupric ammine can be estimated from the color of the solution. The hydroxide ion concentration should be twice as great. The situation is not, however, so simple. The pH of the solution is less than that expected on the basis of Reaction 2, probably because of the acidity of the cupric ammine ion, which contains two water ligands

$$[Cu(NH_3)_4(H_2O)_2]^{2+} + OH^- \rightleftarrows [Cu(NH_3)_4(H_2O)(OH)]^+ + H_2O \qquad (3)$$

The difficulty can be avoided if the solution is buffered to a known pH in the moderately alkaline range. This is accomplished most directly by treating the cupric hydroxide with an equimolar mixture of ammonia and ammonium nitrate

$$Cu(OH)_2(s) + 2NH_3 + 2NH_4^+ \rightleftarrows Cu(NH_3)_4^{2+} + 2H_2O \qquad (4)$$

thus neutralizing the hydroxide ion. The concentrations of ammonia and ammonium ion remain equal, since they are consumed equally, and the pOH of the solution remains buffered at 4.75, the pK_b of the ammonia.

The mass action expression for Equation 4 can be written

$$K = \frac{[Cu(NH_3)_4^{2+}][H_2O]^2}{[Cu(OH)_2]_s[NH_3]^2[NH_4^+]^2}$$

but since the water concentration is essentially constant in water solution, and because cupric hydroxide is a solid, the expression can be written

$$K_4 = \frac{[Cu(NH_3)_4^{2+}]}{[NH_3]^2[NH_4^+]^2}$$

It can be shown that

$$K_4 = \frac{K_f K_{sp}}{K_b^2}$$

by combining the following equilibria with the mass action equation for formation of the cupric ammine complex

$$Cu(OH)_2(s) \rightleftarrows Cu^{2+} + 2OH^-$$

$$K_{sp} = [Cu^{2+}][OH^-]^2$$

$$NH_3 + H_2O \rightleftarrows NH_4^+ + OH^-$$

$$K_b = \frac{[NH_4^+][OH^-]}{[NH_3]}$$

$$\frac{[Cu(NH_3)_4^{2+}]}{[Cu^{2+}][NH_3]^4} \times [Cu^{2+}][OH^-]^2 \times \frac{[NH_3]^2}{[NH_4^+]^2[OH^-]^2}$$

$$= K_f \times K_{sp} \times \frac{1}{K_b^2}$$

Canceling terms gives

$$\frac{[Cu(NH_3)_4^{2+}]}{[NH_3]^2[NH_4^+]^2} = \frac{K_f K_{sp}}{K_b^2}$$

which is the mass action expression for Reaction 4.

Procedure A: Using visual comparison

Prepare four 50 micromole samples of cupric hydroxide as follows. Into each of four clean 13- × 100-mm test tubes place 10 drops of $0.1M$ $CuCl_2$ followed by 12 drops of $0.2M$ NaOH. If you use the same eyedropper, wash it! Stopper the test tubes, shake to mix, and centrifuge. The tubes should be placed in the centrifuge cups so that they counterbalance one another. Otherwise the centrifuge will vibrate wildly. Always balance the head in the centrifuge. Decant the liquid. Fill each tube half full (keep the level the same in each) with distilled water, stopper, shake, and centrifuge. Decant the wash water.

Number the tubes 1, 2, 3, and 4, and in that order add 18, 16, 14, and 12 drops of water followed by 1, 2, 3, and 4 drops of $1M$ NH_3 and 1, 2, 3, and 4 drops of $1M$ NH_4NO_3. Each tube should contain 20 drops (1 ml) total volume. Stopper the tubes, shake vigorously for two minutes, and centrifuge. Observe and record the relative color intensities of cupric ammine (most, second, third, least), and whether a precipitate of cupric hydroxide remains. Discard any tube that does not contain a precipitate.

The estimation of the $Cu(NH_3)_4^{2+}$ ion concentration is accomplished by comparison of color with a solution of known $Cu(NH_3)_4^{2+}$ concentration. Prepare 20 ml of $0.05M$ $CuCl_2$ and reserve for future use. Taking the tube with the least cupric ammine first, decant as much of the solution as possible into a clean 100-mm test tube and add 6 drops of NH_3. Note and record any color change.

Wash the tube containing the precipitate with a little 6M HCl and rinse. Label this the comparison tube, add 6 drops of 6M NH$_3$, about 0.5 ml of water, and then add dropwise the 0.05M CuCl$_2$ until the color is the same as that of the solution from Tube 1. Make the volumes of the two tubes equal before making a final judgment. Observe the color by looking down the length of the tube. Record the number of drops of 0.05M CuCl$_2$ required, and compute the concentration of cupric ammine in the 1 ml of solution of Tube 1. Proceed similarly with the other tubes. However, the color comparison will be easier, if, when over half the copper is in the solution, the analysis is made on the precipitate. Dissolve the precipitate with two drops of 1M NH$_3$NO$_3$, 6 drops of 6M NH$_3$, and about one ml of water.

For each of the tubes in which some cupric hydroxide remained, compute the concentration of ammonia that would have been present if no reaction with cupric hydroxide had occurred from the number of drops of 1M NH$_3$ added. The concentration of NH$_4^+$ ion is the same. Assuming that the stoichiometry is according to Reaction 4, compute the final concentration of ammonia and ammonium ion in each tube. Consequently, the final equilibrium concentration of ammonia or ammonium ion is equal to the initial concentration assuming no reaction, [NH$_3$]$_0$, minus twice the concentration of cupric ammine complex

$$[NH_3]_{eq} = [NH_4^+]_{eq} = [NH_3]_0 - 2[Cu(NH_3)_4^{2+}]$$

For each tube, compute the value of the mass action expression corresponding to Reaction 4 and decide if, within the precision of the experiment, this is a true constant. Combining the average value of the result with the value of the solubility product for cupric hydroxide and the ionization constant of ammonia, compute the value of the formation constant of tetraamminecopper(II), K_f.

Procedure B: Using a photometric colorimeter

The first order decay law appears again in Lambert's[48] and Beer's[49] laws, which relate the light incident on a homogeneous medium to the light transmitted. If the light that is reflected is ignored, the incident light intensity, I_0, equals the intensity absorbed and the intensity transmitted

$$I_0 = I_a + I_t$$

Lambert's (or Bouguer's[50]) law considers only the thickness of the medium. The incident light falling on a film of thickness dl of absorbing substance decreases in intensity $-dI$. The "rate" of decrease is proportional to the intensity

$$-\frac{dI}{dl} = kI$$

Integration gives

$$-\ln I = kl + C$$

C is $\ln I_0$, since $I = I_0$ when $l = 0$

$$-\ln I + \ln I_0 = kl$$

$$\ln \frac{I_0}{I} = kl$$

The constant k was called the extinction coefficient, ϵ, by Bunsen and Roscoe[51], and it is a property of the absorbing material.

Beer's law relates absorption to concentration, c, of the absorbing substance as follows:

$$-\frac{dI}{dc} = k'I$$

Integration gives

$$\ln \frac{I_0}{I} = k'c$$

Combining both laws would give

$$\ln \frac{I_0}{I} = \epsilon l c$$

or

$$2.3 \log \frac{I_0}{I} = \epsilon l c$$

If you do not understand the calculus derivation, do not worry about it. You can utilize the relationship given below even if you do not follow its derivation. The quantity $(I/I_0) \times 100$ is called the *percent transmission*, $\%T$. The quantity $\log(I_0/I)$ is called the *absorbance*, A

$$A = \log \frac{I_0}{I} = \log \frac{100}{\%T} = \frac{\epsilon l c}{2.3}$$

A graph of A versus c should give a straight line if the path length, l, is constant. The slope of the line is $\epsilon l/2.3$. This equation is developed for monochromatic light, that is, light of a single wavelength. Consequently, in practice we usually make a calibration curve, which may deviate from a straight line somewhat, since we may not be using monochromatic light and because of instrument peculiarities.

A photoelectric colorimeter uses a photocell with a galvanometer to measure the transmitted light (Figure 24-3). The intensity transmitted by a solution of absorbing compound is compared to the intensity of light transmitted by the cell containing only solvent (usually water).

The best wavelength to use for measuring the concentration would be the wavelength for which the absorbance can change most for a change in concentration. Consequently, a wavelength near the maximum at 6200 Å or 620 mμ (millimicrons) should be chosen (see Figure 24-2). In the case of colorimeters that use filters, a red filter should be used.

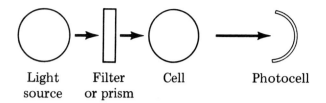

Figure 24-3. Photoelectric colorimeter.

The instructor will demonstrate the operation of the photoelectric colorimeter. Follow instructions. This is an expensive instrument. Do not spill chemicals on the instrument.

Calibration curve

Make solutions that are 0.025, 0.020, 0.010, 0.005, and 0.002M in cupric ammine complex from the volumes listed in Table 24-1 of 0.10M $CuSO_4$, 0.010M $CuSO_4$ (made by 1:10 dilution of 0.10M $CuSO_4$), a solution that is 0.50M NH_3 and 0.50M NH_4NO_3, and water. The NH_3-NH_4NO_3 is made by mixing 10 ml of 1.0M NH_3 with 10 ml of 1.0M NH_4NO_3. Line up five 18- × 150-mm test tubes in a test tube rack. Number them 1 to 5. Using a 5-ml Mohr pipette and a pipette bulb, measure 2.5 ml, 2.0 ml, and 1.0 ml of 0.10M $CuSO_4$ into Tubes 1, 2, and 3 (to use the pipette bulb, see Section 0.5-1). After rinsing the Mohr pipette with 0.010M $CuSO_4$, add the amounts listed in Table 24-1 to Tubes 4 and 5. Rinse the pipette with water and then with NH_3-NH_4NO_3 solution before adding the listed volumes to the tubes. Rinse the pipette with water well, and add the listed volumes of water to the tubes. Shake each tube. Using the colorimeter, measure the absorbance (A) of the samples. If the instrument reads in percent transmission (%T), convert this to A. Make a graph of absorbance versus concentration of copper ammine complex. Draw a smooth curve through the points. The graph should be very nearly a straight line.

Table 24-1

Complex concentration	No. 1, 0.025M	No. 2, 0.020M	No. 3, 0.010M	No. 4, 0.005M	No. 5, 0.002M
0.10M $CuSO_4$	2.5 ml	2.0 ml	1.0 ml	0.0 ml	0.0 ml
0.010M $CuSO_4$	0.0 ml	0.0 ml	0.0 ml	5.0 ml	2.0 ml
0.5M NH_3-0.5M NH_4NO_3	3.0 ml	2.6 ml	1.8 ml	1.4 ml	1.15 ml
Distilled water	4.5 ml	5.4 ml	7.2 ml	3.6 ml	6.85 ml

Measuring the equilibrium

Using a pipette, mix 5 ml of 1.0M aqueous ammonia and 5 ml of 1.0M aqueous ammonium nitrate with 15 ml of distilled water, making a solution that is 0.2M in NH_3 and 0.2M in NH_4NO_3.

Four 0.25-millimole samples of cupric hydroxide are prepared as follows. Into each of four clean 13- × 100-mm test tubes place 2.5 ml of 0.1M $CuCl_2$ followed by 3.0 ml of 0.2M sodium hydroxide. Stopper the test tubes, shake well, and centrifuge. Decant the liquid. Fill each tube half full (keep the level the same in each) with distilled water, stopper, shake, and centrifuge. Decant the wash water. Number the tubes 1, 2, 3, and 4.

Using a 5-ml graduated Mohr pipette, add 3.0 ml of the NH_3-NH_4NO_3 solution, which is $0.2M$ in ammonium nitrate, to Tube 1. Add 2.0 ml of the $0.2M$ solution to Tube 2, 1.0 ml to Tube 3, and 0.5 ml to Tube 4. Then, after rinsing the pipette, add 2.0 ml of distilled water to Tube 1, 3.0 ml to Tube 2, 4.0 ml to Tube 3, and 4.5 ml to Tube 4. Each tube will have a 5-ml volume of solution. The concentrations of ammonia (or ammonium nitrate) in the tubes, assuming no reaction, $[NH_3]_0$, are $0.12M$, $0.08M$, $0.04M$, and $0.02M$, respectively. Cork and shake the tube for two minutes. Then centrifuge the tubes (with the corks on to prevent dirt from falling into them). Observe the color intensities and record the relative intensities of blue color (most, second, third, least), and whether a precipitate of copper hydroxide remains. Discard any tube that does not contain a precipitate.

Decant the solution from a test tube into a tube for use in the photoelectric colorimeter. Measure the absorbance of the solution with the instrument. Do the same with the other solutions. Taking these absorbances (A), read the concentrations of copper ammine complex from your calibration graph.

Calculation

1. Compute what the concentration of ammonia would have been if there had been no reaction with cupric hydroxide, $[NH_3]_0$. The concentration of ammonium ion is the same (given above).

2. Compute the concentration of the cupric ammine complex ion from the colorimeter calibration graph.

3. Assuming the stoichiometry of Reaction 4, compute the concentration of ammonia left. Notice that two ammonia molecules and two ammonium ions are used to form each tetraamminecopper(II) cation. Therefore,

$$[NH_3]_{eq} = [NH_4^+]_{eq} = [NH_3]_0 - 2[Cu(NH_3)_4^{2+}]$$

4. For each tube, compute the value of K_4 and decide if, within the precison of the experiment, it is a constant.

5. Combine the average of the above K_4's with the value of the solubility product of cupric hydroxide and the dissociation constant for ammonia; compute the value of the formation constant of tetraamminecopper(II), K_f.

SUPPLEMENTARY READING

Dickerson, Gray, and Haight (Chapter 10, and Chapter 16, Section 9)

Masterton and Slowinski (Chapter 19)

Sienko and Plane (Chapter 12, Section 6, and Chapters 19 and 22)

Mahan (Chapter 5, Section 7; Chapter 15, Sections 9, 11, and 12; and *University Chemistry* (Chapter 5, Section 7; Chapter 16, Sections 9, 11, and 12)

Brescia, *et al.* (Chapter 25, and Chapter 19, Section 14)

Brown (Chapter 15, Section 10)

Pauling (Chapters 23 and 25)

See especially: F. Basolo and R. Johnson, *Coordination Chemistry* (W. A. Benjamin, New York, 1964).

QUESTIONS

1. When ammonia is added to an aqueous solution of copper ions, what happens? Explain what is observed with a balanced equation.

2. Why don't we measure the equilibrium for Equation 1 directly, instead of measuring that for Equation 4, and then from K_4 calculate the equilibrium constant for Equation 1?

3. According to Beer's law, what relationship exists between the absorbance, A, and the concentration of the absorbing substance in a solution?

PROBLEMS

1. The absorption of light in the visible spectrum for many metal atoms is due to excitation of electrons in d orbitals to higher-energy orbitals. The difference in energy between these orbitals is affected by the arrangement of ligands around the metal ion, what the ligand is, the charge on the metal, and the principal quantum number of the d electrons. It can be seen (Figure 24-2) that the maximum absorption for $Cu(H_2O)_4^{2+}$ is at about 8000 Å, which corresponds to a frequency ($\nu = 1/\lambda$) of 1.25×10^4 cm^{-1}. Because energy is proportional to frequency ($\Delta E = h\nu c$), frequency (or wave number) can be used as a measure of this energy splitting between d orbitals. If there is more than one d electron, the situation is more complicated, but the wavelength of the maximum absorbance still corresponds rather closely to this energy splitting. Calculate the splitting for the $Cu(NH_3)_4^{2+}$ complex ion in wave numbers (cm^{-1}). Does NH_3 as a ligand cause a larger splitting than water? Discuss in terms of crystal field or ligand field theory.

2. Calculate the extinction coefficient at the maximum absorbance (ϵ_{max}) for the tetraamminecopper cation if the cells used in Figure 24-2 were 1 cm in path length (l = 1 cm).

3. Would you expect the tetraamminecopper(II) cation to be diamagnetic or paramagnetic? Discuss.

4. Discuss the kind of orbitals used in bonding by copper in this complex.

EXPERIMENT 25

Equipment and supplies

Six 18- × 150-mm test tubes and corks (or vials with caps)

Two 150-mm or 200-mm test tubes

One 400-ml beaker

Two 125-ml Erlenmeyer flasks

One burner, ring stand, wire gauze

One 50-ml burette

One 5-ml transfer pipette

2.5 g Silver acetate

2.5 g Potassium thiocyanate

1.0 g Silver nitrate

10 ml 0.05M Sodium acetate

10 ml 1.0M Sodium acetate

10 ml 0.10M Ammonia

10 ml 0.10M Nitric acid

10 ml 1.0M Sodium nitrate

10 ml Saturated ferric ammonium sulfate

30 ml 6N Nitric acid

6 Filter papers

Time requirement

One-half hour for preparing solutions.

Two hours for analysis of the solutions.

THE SOLUBILITY OF SILVER ACETATE

The purpose of this experiment is to determine the solubility product constant of silver acetate by measuring the solubility.

INTRODUCTION

In this experiment, saturated solutions of silver acetate will be analyzed for silver ion by the Volhard[52] titration. In the Volhard method, the silver solution is titrated with standardized potassium thiocyanate solution. Silver thiocyanate precipitates

$$Ag^+ + SCN^- \rightarrow AgSCN\downarrow$$

Ferric ion is used as an indicator. When all the silver ion is precipitated, excess thiocyanate will be present in the solution. The thiocyanate forms a deep red complex ion with ferric ion

$$Fe^{3+} + SCN^- \rightarrow Fe(SCN)^{2+}$$

The first persistent pink color is the end point. The amount of silver ion found in a solution is the solubility of silver acetate in the solution.

Silver acetate is a slightly soluble salt. The equilibrium between solid silver acetate and dissolved silver acetate in a saturated solution can be described as

$$AgOAc(s) \rightleftarrows Ag^+ + {}^-OAc$$

where the dissolved ionic salt is present as dissociated ions, Ag^+ and ^-OAc. The solubility product is written

$$K_{sp} = [Ag^+][^-OAc]$$

When silver acetate only is dissolved in water, the silver ion concentration equals that of the acetate ion, and this concentration is equal to the molar solubility, s, of the silver acetate (moles of silver ion per liter)

$$[Ag^+] = [^-OAc] = s$$

Therefore,

$$K_{sp} = s^2$$

If there are added soluble acetate salts, such as sodium acetate, then the silver ion concentration is not equal to the concentration of acetate ion. The molar solubility, s, is still the silver ion concentration found by titration. The acetate ion concentration is the concentration of the added acetate, C, plus the acetate ion from dissolved silver acetate, which is s, the solubility of silver acetate

$$[^-OAc] = C_{NaOAc} + s$$

Then the solubility product equation in the presence of the common ion, acetate ion, becomes

$$K_{sp} = s(C_{NaOAc} + s)$$

Ionic equilibria are affected by the concentration of ions in the solution. Strictly speaking, the solubility product equation should be written

$$K_{sp} = (a_{Ag+})(a_{-OAc})$$

where a is the activity of the respective ions. Concentration is a more convenient quantity to measure than activity, so a relationship between activity and concentration is defined as

$$a_{Ag+} = \gamma[Ag^+]$$

where γ is the "activity coefficient." Activity is an "effective concentration." When a solution is extremely dilute, γ approaches unity, and, consequently, activity equals concentration. But in less dilute solutions, γ is less than one, which means that the effective concentration, the activity, of ion is less than the measured concentration. As a result, the *solubility product* of silver acetate (calculated from K_{sp} = $[Ag^+][^-OAc]$ using the respective concentrations) in a $1M$ sodium acetate solution, in which the concentration of ions is about $2M$, will be greater than the solubility product would be in water, in which only about $0.1M$ dissolved silver and acetate ions are present.

PROCEDURE

Clean six test tubes and corks (or vials and caps). Number them 1, 2, 3, 4, 5, 6.

Add about 0.3 g of solid silver acetate to Tubes 1 through 4. Add about 0.5 g to Tubes 5 and 6. To Tube 1, add 10 ml of distilled water. To Tube 2, add 10 ml of $0.05M$ sodium acetate. To Tube 3, add 10 ml of $1.0M$ sodium acetate. To Tube 4, add 10 ml of $1.0M$ sodium nitrate. To Tube 5, add 10 ml of $0.10M$ ammonia. To Tube 6, add 10 ml of $0.10M$ nitric acid. Close all tubes (or vials) tightly. Heat water in a beaker to $50°C$. Warm the tubes in this bath for five minutes, then shake them for a few minutes. Warm them again and shake for five minutes. Let the tubes stand for a week, shaking them occasionally, if possible.

Standardization of KSCN

Weigh approximately enough potassium thiocyanate to make 250 ml of 0.1M KSCN. Dissolve it completely in distilled water and stopper the flask. This KSCN solution now will be standardized against solid silver nitrate, which is available as a pure reagent.

Weigh two samples of about 0.5 g of silver nitrate on pieces of smooth paper on a triple beam balance. (*Warning*: Silver nitrate and other silver compounds are poisonous, and also can cause blisters. Silver compounds turn black in light. Clean up immediately if any compound is spilled.) Weigh a flask accurately on an analytical balance. Add one of the 0.5-g portions of silver nitrate to the flask and reweigh accurately. The difference in weights is the weight of the silver nitrate sample. Do the same with the other portion. Add 20 ml of distilled water (NOT tap water) and 1 ml of saturated ferric ammonium sulfate solution to each flask.

Add 20 ml of distilled water and 1 ml of indicator to a third flask. Rinse and fill a burette with the KSCN solution. Titrate the distilled water, drop-by-drop, until a faint red color persists. The volume of KSCN solution added is the "blank" to be subtracted from all later titrations. This blank is the amount of KSCN solution required to make an observable red color. Titrate the silver nitrate samples. Calculate the molar concentration of KSCN in the solution for each sample. At the end point, the number of moles of KSCN equals the number of moles of silver in the flask

$$\text{moles of KSCN} = \text{moles of Ag}^+$$

$$(V_{KSCN} - V_{blank})[KSCN] = \frac{w}{\text{mol wt}}$$

where V is volume; w, the weight of $AgNO_3$; and mol wt, the molecular weight of $AgNO_3$. Average the two results obtained for the concentration of the KSCN solution.

Analysis of silver acetate solutions

Filter the contents of each tube into a clean test tube, using a fluted filter paper. Pipette 5 ml of each solution into a 125-ml Erlenmeyer flask. To each flask, add 5 ml of 6N nitric acid, using a graduated cylinder. Add 10 ml of water to each flask and 1 ml of ferric indicator solution. Titrate each solution with the standard potassium thiocyanate. Calculate the concentration of silver ion in each of the samples, realizing that at equivalence

$$\text{moles of KSCN} = \text{moles of Ag}^+$$

$$(V_{KSCN} - V_{blank})[KSCN] = 5 \text{ ml}[Ag^+]$$

and that 5.0 ml is the volume of the silver solution analyzed. The

[Ag⁺] calculated is the solubility of silver acetate for that solution, s. Calculate the solubility product, K_{sp}, for Tubes 1, 2, 3, and 4. Calculate only the solubility of silver acetate in Tubes 5 and 6.

Discuss the significance of your results. Explain the differences in solubility between the contents of the test tubes.

SUPPLEMENTARY READING

Dickerson, Gray, and Haight (Chapter 17)

Masterton and Slowinski (Chapter 16)

Sienko and Plane (Chapter 12, Section 7)

Mahan (Chapter 6, Section 1)

Brescia, *et al.* (Chapter 19, Section 12)

Brown (Chapter 15, Section 11)

Hammond, Osteryoung, Crawford, and Gray (Chapter 9)

QUESTIONS

1. Is the solubility of silver acetate larger, or smaller, in the presence of a soluble acetate salt such as sodium acetate?
2. Is the solubility of silver acetate affected by the presence of ammonia? Write a balanced equation that explains this phenomenon.
3. Is the solubility of silver acetate affected by the presence of nitric acid? Write an equation that accounts for this effect.
4. How does the solubility product change with the concentration of ions in the solution? Suggest a reason for this behavior.

PROBLEMS

1. Calculate the solubility product for the slightly soluble salt lead chloride ($PbCl_2$) from the solubility given in a handbook, which is 0.99 g of $PbCl_2$ in 100 ml of water at 20°C. Calculate the solubility of $PbCl_2$ in 1.0M NaCl solution, using the solubility product calculated above.

2. The solubility of hexaamminecobalt(III) chloride (see Experiment 15) is 7.00 g per 100 ml of water at 20°C. Estimate the solubility of this salt in dilute (perhaps 3.0M) hydrochloric acid solution. Notice that in the preparation of this compound (Experiment 15), hydrochloric acid was used in this way to decrease the solubility of the salt.

3. What minimum concentration of aqueous ammonia would be needed to dissolve 1.00 g of silver chloride and keep it in solution. The volume of the solution is to be 200 ml. The overall reaction is

$$AgCl(s) + 2NH_3 \rightarrow Ag(NH_3)_2^+ + Cl^-$$

The solubility product of silver chloride is 1.7×10^{-10}, and the dissociation constant for $Ag(NH_3)_2^+$ is 6.0×10^{-8}.

4. Sodium thiosulfate ("hypo") is used as a fixer for photographic films and papers. The thiosulfate ion complexes with silver salts that are unexposed to light, and thus dissolves them. The exposed silver salts were "developed" previously, and at this "fixing" stage are metallic silver, which does not complex. If it is desired that a given fixing bath be capable of dissolving 40 g of silver chloride per liter, what must be (a) the concentration of free thiosulfate ion when the solution has dissolved the 40 g of silver chloride per liter, and (b) the initial concentration of thiosulfate ion before reaction with the silver chloride? The overall reaction to be considered is

$$AgCl(s) + 2S_2O_3^{2-} \rightarrow [Ag(S_2O_3)_2]^{3-} + Cl^-$$

The solubility product for silver chloride is 1.7×10^{-10}, and the dissociation constant for the thiosulfatoargentate anion is 1×10^{-13}.

EXPERIMENT 26

Equipment and supplies

For this experiment it is efficient to have the known and reagent solutions in dropper bottles or plastic squeeze bottles.

Six 13- × 100-mm test tubes

Medicine dropper

Centrifuge with head to take 100-mm test tubes

Beaker

Burner, wire gauze, ring stand

Test tube holder

Spatula

Stirring rod

Plastic squeeze bottle

10-25 ml Unknown or unknowns

2 ml 0.02M Silver nitrate

1 ml 0.05M Lead nitrate

3 ml 6M Hydrochloric acid

0.5 ml 0.5M Potassium chromate

15 ml 6M Ammonia

0.5 ml 6M Nitric acid

1 ml 0.02M Mercuric nitrate

4 ml 0.02M Ferric chloride

3 ml 0.02M Nickel nitrate

1 ml 0.02M Copper nitrate

2 ml 0.02M Calcium chloride

1 ml 0.02M Arsenic trichloride

2 ml 1M Ammonium nitrate

1 ml 0.05M Aluminum choride

2 ml Dimethylglyoxime solution

5 ml 6M Sodium hydroxide

1 ml 0.1M Potassium thiocyanate

1 g Ammonium chloride (solid)

1 ml Potassium ferrocyanide solution

10 ml 1M Ammonium carbonate

1 ml 0.01M Ammonium chloride

1 ml 1M Barium nitrate

5 g Ferrous sulfate (solid)

10 ml Concentrated sulfuric acid

Nine pound cylinder of gaseous hydrogen sulfide for 300 students

Litmus paper

100 ml Water sample

1 ml each Buffer solutions pH 6, 6.5, 7, 7.5, 8

1 ml or less of Bromthymol blue indicator

Time requirement

Three to four three-hour laboratory periods

INORGANIC QUALITATIVE ANALYSIS

The purpose of these experiments is to present the elementary operations of qualitative analysis, and thus acquaint the student with some reactions of the more common metals.

INTRODUCTION

In qualitative analysis, the *presence* of a substance is detected, in contrast to quantitative analysis where the *amount* present is determined. In these experiments, the number of different cations to be detected will be limited to ten: Ag^+, Pb^{2+}, Al^{3+}, Fe^{3+}, Hg^{2+}, As^{3+}, Ni^{2+}, Cu^{2+}, NH_4^+, and Ca^{2+}. These can be separated from one another because of differences in their chemical properties. You will receive one or more unknowns that will contain any or all of these ions, and you will analyze the unknown solution for them. Be careful not to contaminate your unknown; never pour *anything* into the unknown, and do not pour any of the unknown back into your supply either!

The first separation is based on the difference in the solubilities of the chlorides. Some of the ions (Pb^{2+} and Ag^+) form insoluble chlorides. These are called Group I. The second separation will be based on different solubilities of the sulfides in acidic solution. As seen in Table 26-1, the solubility products of the sulfides of Hg, Al, Fe, As, Ni, and Cu are very small. If the concentration of any metal ion in the solution to be analyzed is $0.02M$, then in order not to precipitate nickel ions, the sulfide ion concentration must be smaller than $1.5 \times 10^{-19} M$

$$NiS \rightleftarrows Ni^{2+} + S^{2-}$$

$$[Ni^{2+}][S^{2-}] = 3 \times 10^{-21}$$

$$[S^{2-}] = \frac{3 \times 10^{-21}}{2 \times 10^{-2}} = 1.5 \times 10^{-19} M$$

The sulfide ion concentration can be controlled by adjusting the hydronium ion concentration, because of the following equilibria

$$H_2S \overset{K_1}{\rightleftarrows} H^+ + HS^-$$

$$HS^- \overset{K_2}{\rightleftarrows} H^+ + S^{2-}$$

Table 26-1. Solubility products and solubilities near room temperature (Solubilities in grams per 100 ml of water are in parentheses; s denotes soluble; v.s., very soluble; i, insoluble; d, decomposes.)

Cation[a]	Sulfides	Chlorides	Hydroxides	Carbonates	Sulfates
[a]NH_4^+	v.s.	(29.7)	s	(100)	(76)
Mg^{2+}	d	(54.5)	1×10^{-11}	3×10^{-5}	(26)
K^+	s	(34.7)	(107)	(112)	(12)
Ba^{2+}	s	(31)	(5.6)	8×10^{-9}	1×10^{-10}
Sr^{2+}	s.d.	(43.5)	3×10^{-4}	1.6×10^{-9}	4×10^{-7}
[a]Al^{3+}	d	(70)	3.7×10^{-15}	—	(31)
Na^+	(30)	(35.7)	(42)	(7)	(5)
[a]Ca^{2+}	1×10^{-6}	(59.5)	8×10^{-6}	8.7×10^{-9}	2×10^{-4}
Mn^{2+}	7×10^{-16}	(62)	4×10^{-14}	3×10^{-9}	(52)
[a]Fe^{2+}	4×10^{-19}	(64)	1×10^{-14}	3×10^{-9}	(8.5)
[a]Fe^{3+}	—	(75)	1×10^{-36}	—	—
[a]Ni^{2+}	3×10^{-21}	(64)	4×10^{-15}	6×10^{-9}	(29)
Co^{2+}	5×10^{-22}	(45)	1.6×10^{-16}	i	(36)
Zn^{2+}	1×10^{-22}	(43)	1.8×10^{-14}	6×10^{-11}	(86)
Sn^{2+}	1×10^{-26}	(84 d)	—	—	(33)
Cd^{2+}	1×10^{-28}	(140)	2×10^{-17}	i	(75)
[a]Pb^{2+}	7×10^{-29}	1.6×10^{-5}	4×10^{-15}	3×10^{-14}	1×10^{-8}
[a]As^{3+}	1×10^{-33}	d	—	—	—
[a]Cu^{2+}	8×10^{-37}	(71)	10^{-20}	i	(14)
[a]Ag^+	5.5×10^{-51}	1.6×10^{-10}	1.5×10^{-8}	6×10^{-12}	2.4×10^{-8}
[a]Hg^{2+}	1.6×10^{-54}	(7)	—	—	1×10^{-10}

[a] These cations are treated in Experiment 26

Combining both equilibria in one mass action expression gives

$$H_2S \overset{K_1 K_2}{\rightleftharpoons} 2H^+ + S^{2-}$$

$$(1.1 \times 10^{-7})(1 \times 10^{-14}) = K_1 K_2 = \frac{[H^+]^2 [S^{2-}]}{[H_2S]}$$

Since saturated hydrogen sulfide solutions are $0.1M$, the hydronium ion concentration must be about $0.027M$, if the sulfide ion is not to exceed $1.5 \times 10^{-19}M$.

$$1.1 \times 10^{-21} = [H^+]^2 \frac{1.5 \times 10^{-19}}{0.1}$$

$$[H^+]^2 = \frac{1.1 \times 10^{-22}}{1.5 \times 10^{-19}} = 7.3 \times 10^{-4}$$

$$[H^+] = 2.7 \times 10^{-2} = 0.027M$$

$$pH = 1.6$$

If the hydronium ion concentration is maintained at this value, all metal sulfides of smaller solubility products than NiS will precipitate, but NiS, and those that are more soluble, will not. The [H$^+$] must be lower than $0.027M$ before NiS will precipitate. The cations precipitated with hydrogen sulfide in acidic solution are called Group II.

If the [H$^+$] is decreased by buffering the solution with ammonia so that the amounts of ammonium chloride and ammonia are about equal (about [H$^+$] = 10^{-8}), then the more soluble sulfides can be precipitated. In our selection of cations, Fe^{2+} and Ni^{2+} now would be precipitated as sulfides and would comprise Group III. Al^{3+} precipitates as the hydroxide when ammonia is added. Aluminum sulfide is not stable in water.

Cations whose sulfides are soluble, such as Ba^{2+}, Sr^{2+}, Ca^{2+}, Na$^+$, K$^+$, NH$_4^+$, and Mg^{2+} still will be in solution. The filtrate is acidified and the hydrogen sulfide boiled out. Then the solution is buffered with ammonia (and ammonium chloride), and Ba^{2+}, Sr^{2+}, and Ca^{2+} are precipitated as carbonates by adding ammonium carbonate.

Table 26-2 summarizes the separation scheme for the cations.

Precipitation can occur only when the product of the concentrations exceeds the solubility product. However, there are reasons why precipitation might not be observed even then: (1) The amount of material must be sufficient to be seen, (2) The rate of precipitation may be too slow, that is, supersaturation of the solution may occur, (3) Furthermore, activities of the ions may be much lower than the actual concentrations, and, consequently, calculated solubilities may be in considerable error.

PROCEDURE

In the tests and procedures that follow, write an equation for every reaction and also indicate the result. For example,

$$AgNO_3 + HCl \rightarrow AgCl\downarrow + HNO_3$$
white
precipitate

The results of tests should be written as soon as they are done, NOT LATER. The appropriate time to write equations will be indicated for Group I, but not in later sections.

Part 1: Group I, Ag$^+$, Pb^{2+}

Measure 1 ml of water into a 13- × 100-mm test tube. Look at the amount in the test tube and fix this approximate amount in your mind.

Step 1

Pour estimated 1-ml samples of $0.02M$ AgNO$_3$ and $0.05M$ Pb(NO$_3$)$_2$ into two clean test tubes. "Clean test tube" in all these experiments means washed clean to the eye and rinsed three times with distilled water. Add dilute ($1M$) HCl with gentle shaking until precipitation is complete in each tube (2 drops of $6M$ HCl should be enough). Write equations for both reactions.

Table 26-2. Separation of cations in the scheme of qualitative analysis (NH_4^+ is analyzed separately)

Ag^+, Pb^{2+}, Cu^{2+}, As^{3+}, Hg^{2+}, Fe^{2+}, Fe^{3+}, Ni^{2+}, Al^{3+}, Ca^{2+} Add HCl						
Ppt. AgCl, $PbCl_2$ Add hot water	Soln. Cu^{2+}, As^{3+}, Hg^{2+}, Fe^{2+}, Fe^{3+}, Ni^{2+}, Al^{3+}, Ca^{2+} Saturate with H_2S (acidic solution)					
Ppt. AgCl	Soln. Pb^{2+}	Ppt. CuS, As_2S_3, HgS, (S) Add NH_3, saturate with H_2S		Soln Fe^{2+}, Ni^{2+}, Al^{3+}, Ca^{2+} Add NH_3, NH_4^+, saturate with H_2S		
		Ppt. CuS, HgS Add hot HNO_3	Soln. AsS_3^{3-}, (S_2^{2-})	Ppt. FeS, NiS, $Al(OH)_3$ Add HCl	Soln. Ca^{2+}	
		Ppt. HgS	Soln. Cu^{2+}		Ppt. NiS	Soln. Fe^{2+}, Al^{3+} Add O_2, NaOH
					Ppt. $Fe(OH)_3$	Soln. $Al(OH)_4^-$

Step 2

Centrifuge both tubes after corking them. Centrifugation for a minute or so should be sufficient. Decant the liquid. Add 1 ml of distilled water to each tube and heat both tubes in a beaker of boiling water, shaking them occasionally. Write equations.

Step 3

To the clear hot solution of lead chloride, add a few drops of 0.5M K_2CrO_4 (potassium chromate) solution. Write an equation. To 1 ml of 0.02M silver nitrate, add the same amount of potassium chromate. Write an equation (specify the color of the precipitate).

Step 4

To the silver chloride precipitated in Step 2, add a little 6N ammonia solution. Write an equation. What happens? Acidify this solution with a drop or two of 6N HNO_3 from an eyedropper. Write an equation.

Step 5. Analysis of the unknown

a) Place 1 ml of your unknown solution in a test tube. Add 1 drop of 6M hydrochloric acid. Let the precipitate (if any) settle, then add another drop of 6M HCl to test completeness of precipitation. Centrifuge, and pour off the liquid. The precipitate contains Group I cations; the liquid contains Groups II-IV. The liquid can be discarded this time.

b) Add 2 ml of distilled water to the precipitate and heat it in a bath of boiling water. Shake or stir occasionally. Decant the clear hot liquid. Test for Pb^{2+} as in Step 3. Test the precipitate (if any is left) for Ag^+ as in Step 4. Write equations for each step, and state your conclusions. State in a sentence the ions you have found in your unknown.

Part 2: Group II, Hg^{2+}, (Pb^{2+}), Cu^{2+}, As^{3+}

The Group II cations are precipitated with hydrogen sulfide in acidic solution. In this experiment, the acid concentration need not be accurately controlled because Co, Zn, Sn, and Cd will not be present in the unknowns. The [H^+] must be 0.027M or greater. Hydrogen sulfide is a gas with a very unpleasant odor, and is EXTREMELY POISONOUS. Hydrogen sulfide is particularly insidious, since the sense of smell may become fatigued and fail to give warning of high concentrations. The hydrogen sulfide generator or tank should be kept in the back of the hood. The instructor should check to see that the hood is operating. Small amounts of hydrogen sulfide should be boiled out of water solutions in the hood, if possible. If hood space is insufficient, windows should be kept open for good ventilation. When the hood is in use, the glass hood door should be kept half-way closed. Never put your head inside the hood with hydrogen sulfide.

Step 1

a) In separate numbered test tubes, obtain 1-ml samples of 0.02M solutions of $Hg(NO_3)_2$, $FeCl_3$, $Ni(NO_3)_2$, $Cu(NO_3)_2$, $CaCl_2$, and $AsCl_3$. To each, add one drop of 6N HCl, and saturate with H_2S by bubbling the gas through the solution. In the reaction with ferric ion, sulfide is oxidized to sulfur, which is the milky white precipitate

$$2Fe^{3+} + S^{2-} \rightarrow 2Fe^{2+} + S\downarrow$$

Write equations for all reactions.

b) Add 6 drops of $6N$ ammonia to the tubes that contain iron, nickel, or calcium ions. Centrifuge and discard the solution. Add 1 ml of $1M$ HCl to each (3 drops $6N$ HCl + 17 drops H_2O).

c) The behavior of Pb^{2+} and Ag^+ in the presence of hydrogen sulfide may be tested on the $0.02M$ solutions of $AgNO_3$ and $Pb(NO_3)_2$.

Step 2

Heat the test tubes in a hot water bath for five minutes or longer. Centrifuge the test tubes containing HgS, CuS, PbS, and As_2S_3. Decant the liquid. Add 10 drops of $6M$ ammonia and 10 drops of water to each test tube, and saturate with H_2S. Heat in a hot water bath. Stir with a stirring rod to break up the precipitate. The arsenic sulfide dissolves in excess sulfide ion (as do Sb_2S_3 and SnS_2) forming a complex ion, trisulfidoarsenite(III)

$$As_2S_3\downarrow + 3S^{2-} \rightarrow 2AsS_3^{3-}$$

Decant, if necessary, and carefully add 12 drops of $6M$ HCl, or enough to acidify the solution of arsenic

$$2AsS_3^{3-} + 6H^+ \rightarrow As_2S_3\downarrow + 3H_2S$$

If there is a tannish-white precipitate of sulfur present, it also will dissolve in excess sulfide ion, forming disulfide ions

$$S + S^{2-} \rightarrow S_2^{2-}$$

$$S_2^{2-} + 2H^+ \rightarrow H_2S + S\downarrow$$

Step 3

Centrifuge the test tubes containing precipitates of HgS, CuS, and PbS as in Step 2. Decant the liquid. Wash the precipitates by adding a few drops of ammonium nitrate. Centrifuge and pour off the liquid. Add six drops of water and nine drops of $6M$ nitric acid to each test tube and heat in a bath of boiling water. The nitric acid oxidizes sulfide ion to sulfur. Removing sulfide ion shifts the equilibrium between metal ion and metal sulfide, and the metal ion goes into solution

$$CuS(s) \rightleftarrows Cu^{2+} + S^{2-}$$

$$8H^+ + 3S^{2-} + 2NO_3^- \rightarrow 3S + 2NO + 4H_2O$$

Both the PbS and CuS react, but the solubility product of mercuric sulfide is so small (see Table 26-1) that it does not. Centrifuge the test tube containing cupric ion and pour the solution into a clean test tube. Carefully add 15 drops of $6N$ ammonia to the test tube containing the cupric ions. A deep blue color indicates copper.

Part 3: Group III, Fe^{2+}, Ni^{2+}, Al^{3+}

After the Groups I and II ions have been removed, Group III ions will be precipitated by hydrogen sulfide in slightly basic solution. The sulfide of aluminum cannot be precipitated from water, but the gelatinous aluminum hydroxide is precipitated with ferrous or ferric hydroxides, which are converted to ferrous sulfide by hydrogen sulfide. This leaves only Group IV, of which Ca^{2+} and NH_4^+ are members. (In more complete treatments, this fourth group would be broken into Group IV, Ca^{2+}, Sr^{2+}, Ba^{2+}, and Group V, Na^+, K^+, NH_4^+, Mg^{2+}.)

Step 1

Put 1-ml portions of $0.02M$ $FeCl_3$, $0.05M$ $AlCl_3$, and $0.02M$ $Ni(NO_3)_2$ in clean test tubes. Add $6N$ ammonia dropwise to each. The nickel hydroxide precipitate eventually will dissolve, because tetraamminenickel(II) cation is formed. Notice the colors.

Step 2

Add a few drops of dimethylglyoxime solution to 1 ml of $0.02M$ $Ni(NO_3)_2$. The structure of the nickel complex is shown in Figure 26-1.

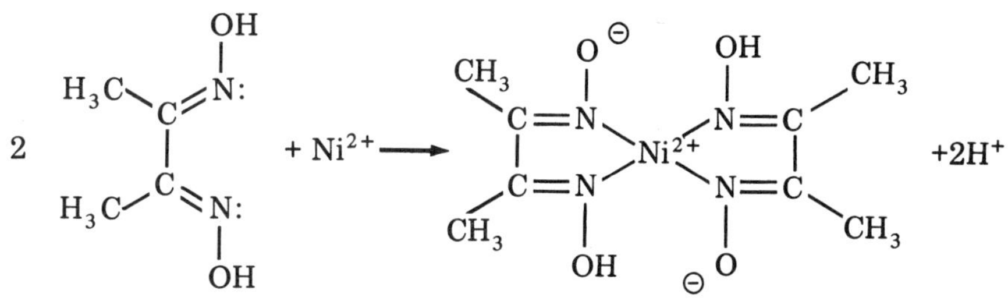

Figure 26-1. Reaction of dimethylglyoxime and nickel ion.

Step 3

Add $1M$ NaOH dropwise to 1-ml portions of $0.02M$ $FeCl_3$, $0.02M$ $Ni(NO_3)_2$, and $0.05M$ $AlCl_3$. The aluminum hydroxide precipitate will dissolve eventually, because of the formation of the aluminate ion

$$Al(OH)_3 + OH^- \rightarrow Al(OH)_4^-$$

Step 4

To 1 ml of $FeCl_3$ solution add a drop of $0.1M$ KSCN.

Part 4: Analysis of the unknown for Groups II and III

Step 1

To 1 ml of the unknown solution, add two drops of $6M$ HCl. Stir for a few minutes and then centrifuge. Decant the liquid into a clean test tube. The precipitate is Group I.

Step 2

a) Saturate the solution with H_2S. Heat the test tube in a beaker of boiling water for 10 minutes. Centrifuge. The precipitate contains Group II. Decant the liquid into a clean test tube.

b) Treat the precipitate containing Group II with 10 drops of 6M aqueous ammonia and 10 drops of water. Saturate with H_2S. Stir the precipitate and heat in a beaker of boiling water. Centrifuge. Decant the solutions into a clean tube. The solution contains AsS_3^{3-} (if present). Carefully add 12 drops of 6M HCl, or enough to acidify the solution.

c) To the residue containing HgS and CuS (and possibly a trace of PbS) from Step 2b, add 10 drops of 6M HNO_3 and five drops of water. Heat the test tube in boiling water for a few minutes. Centrifuge and decant the solution into a clean test tube. To the solution, add excess ammonia to test for Cu^{2+}. A black residue after oxidation (white S also will be present) is HgS.

Step 3

a) To the solution produced from Step 2a, add six drops of 6M ammonia. Test with litmus to see that the solution is basic. Saturate with H_2S. Heat the test tube for 10 minutes in a beaker of boiling water. Centrifuge. Decant the liquid into a clean test tube. This solution will contain Ca^{2+} ion, if it was present. The precipitate contains Group III, Fe^{3+}, Fe^{2+}, Ni^{2+}, and Al^{3+}.

b) To the precipitate produced from Step 3a, add five drops of 6M HCl and 25 drops of water. Cork the tube and shake with air to oxidize ferrous ions to ferric ions

$$4H^+ + 4Fe^{2+} + O_2 \rightarrow 4Fe^{3+} + 2H_2O$$

Stir and mix for 3-5 minutes. Centrifuge. Decant the solution into another clean test tube.

c) If a black residue remains (indicating the probable presence of nickel), add a few drops of concentrated nitric acid to it and heat to boiling in a water bath. Centrifuge, then decant the solution into a clean test tube. Neutralize the filtrate with 6M ammonia and add excess ammonia. Centrifuge, if necessary. Add two drops of dimethylglyoxime solution. A red precipitate indicates nickel.

d) Boil the solution from Step 3b. Neutralize the clear solution with 1M sodium hydroxide dropwise, then add an excess. A brown precipitate indicates iron. Centrifuge, and decant the solution into a clean test tube. Add solid ammonium chloride to the solution to reduce the pH to that of the NH_4^+-NH_3 buffer. Be sure that the solution is buffered. Check the pH. A gelatinous white precipitate probably is aluminum hydroxide.

e) The presence of iron can be confirmed as follows. Dissolve the precipitate of iron hydroxide in a little dilute HCl. To the solution, add about 1 ml of potassium ferrocyanide, $K_4Fe(CN)_6$. A precipitate of Prussian blue, possibly $KFe[Fe(CN)_6] \cdot H_2O$, confirms the presence of iron.

This compound has an interesting structure, as shown in Figure 26-2. The iron atoms are arranged at the corners of a cube with cyanide ions between. Each iron is attached to six cyanide ions, only three of which are shown. The other three extend into adjoining cubes. The iron

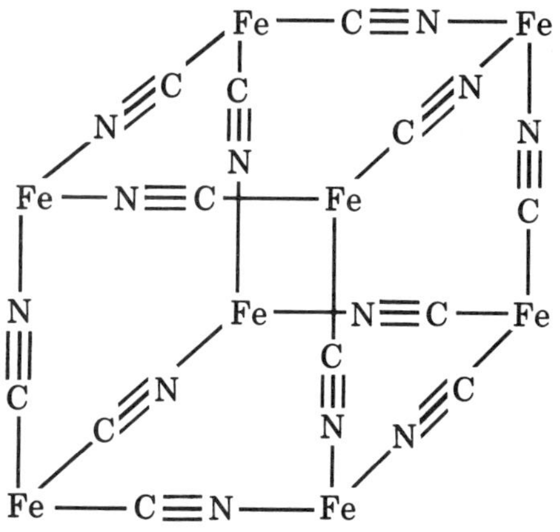

Figure 26-2. Prussian blue lattice.

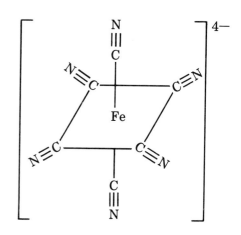

Figure 26-3. Ferrocyanide complex ion, hexacyanoferrate(II).

atoms derived from the potassium ferrocyanide, $K_4[Fe(CN)_6]$ (Figure 26-3), are those that are attached to six carbon atoms. The ferric iron atoms are those attached to six nitrogens. The potassium ions are located in the middle of every other cube in the framework. A water molecule fills the middle of the other cubes.

Part 5: Calcium and ammonium ions

The alkaline earth ions Mg^{2+}, Ca^{2+}, Sr^{2+}, and Ba^{2+} can be separated only by the differences in the solubility of their carbonates, chromates, sulfates, or oxalates. Addition of carbonate ion in concentration less than $10^{-3} M$ should precipitate Ca^{2+}, Sr^{2+}, and Ba^{2+} (see Table 26-2). Ammonium ion must be detected in a fresh sample of unknown solution, because ammonia has been added in the steps described previously.

Step 1

To 1 ml of $0.01M$ calcium chloride solution, add $1M$ ammonium carbonate dropwise.

Step 2

To 1 ml of $0.01M$ ammonium chloride, add $6M$ NaOH dropwise. Hold a moistened red litmus paper in the mouth of the test tube without touching the sides. A blue color indicates ammonia. Gentle warming may be necessary. The color may be faint.

$$NH_4^+ + OH^- \rightarrow H_2O + NH_3\uparrow$$

Very cautiously smell the solution to detect the characteristic ammonia smell. Perform this test on 1 ml of unknown solution. If copper and nickel ions are present, the test for ammonia will be faint because these ions complex strongly with ammonia, forming, for example, $[Cu(NH_3)_4]^{2+}$, and $[Ni(NH_3)_6]^{2+}$.

Step 3

Acidify the clear solution produced from Part 4, Step 3a, from which Groups I, II, and III have been removed. Boil the solution over a flame to remove hydrogen sulfide. Make the solution basic by adding $1M$ sodium hydroxide dropwise. If too much is added, $Ca(OH)_2$ may precipitate. Then add $1M$ ammonium carbonate dropwise. A white precipitate is calcium carbonate.

Part 6: Anions, Cl^-, SO_4^{2-}, NO_3^-, CO_3^{2-}, HCO_3^-

Only the five anions listed will be present in the unknown.

Step 1

Addition of dilute sulfuric acid to the unknown solution will convert carbonate or hydrogen carbonate (bicarbonate) to acid forms

$$CO_3^{2-} + 2H^+ \rightarrow H_2O + CO_2$$

$$HCO_3^- + H^+ \rightarrow H_2O + CO_2$$

The carbonate and bicarbonate solutions will give off bubbles of carbon dioxide that can be detected with wet blue litmus in the mouth of the test tube.

Step 2

If a solution containing chloride ion is acidified with nitric acid and treated with silver nitrate, a precipitate of silver chloride will form. Although silver sulfate also is insoluble, it will not precipitate in acidic solution. Silver carbonate is also only slightly soluble (see Table 26-1), but it is yellow. Carbonate ion can be removed by acidifying the solution with nitric acid and boiling it for a minute or two.

Step 3

Barium nitrate solution can be used to detect sulfate ion by the formation of a white precipitate of barium sulfate. Barium carbonate is also only slightly soluble; consequently, carbonate must be removed first (if present). Carbonate can be removed by acidifying the solution with $6M$ HNO_3 or HCl, and boiling for a minute or two. Then test for sulfate. If a test for Cl^- is desired, the acid used must be $6M$ nitric acid.

Step 4

To 1 ml of a known nitrate solution (or the unknown) that has been made acidic with dilute sulfuric acid, add 1 ml of saturated ferrous sulfate solution, which can be made from solid ferrous sulfate. Incline the test tube at roughly $45°$ and add about 1 ml of concentrated sulfuric acid slowly, letting it run down the side of the test tube. It is important to avoid mixing the sulfuric acid with the lower layer as much as possible. A brown ring forms at the interface between the two layers. This brown color is due to formation of the ferrous nitrosyl complex, $Fe(NO)^{2+}$

$$NO_3^- + 3Fe^{2+} + 4H^+ \rightarrow 3Fe^{3+} + NO + 2H_2O$$

$$Fe^{2+} + NO \rightarrow Fe(NO)^{2+}$$

Be *very careful* when using concentrated sulfuric acid. Do not add water to concentrated sulfuric acid. A large amount of heat will be liberated that will cause the water or solution to boil, possibly spattering sulfuric acid on you or your surroundings. Sulfuric acid is corrosive, very much so when hot.

Part 7: Qualitative analysis of water

Determine the pH of the water sample using the techniques of Experiment 23, Indicators and pH. Since tap water will be only slightly acidic or slightly basic, test it over the range pH 6-pH 8 using buffer solutions of pH 6, 6.5, 7, 7.5, and 8 as comparison standards, with the indicator bromthymol blue. Compare colors of solutions made up uniformly, and estimate the pH to 0.2 of a pH unit. Incidentally, note that if the water is even slightly basic, cations with very insoluble hydroxides (see Table 26-1) cannot be present. Basic water is desirable in this respect, because many such metal cations are bad for health (e.g., Pb^{2+} and Cd^{2+}).

Perform a qualitative analysis on the tap water, using the scheme in Parts 1-6. Because the concentrations of ions in the tap water are likely to be small, less than $0.005M$, the water sample should be concentrated before analysis. Boil 100 ml of the water sample down to just under 10 ml in a 50-ml beaker, as described in Experiment 3. Pour the concentrated sample into a graduated cylinder, and add distilled water until the volume is some convenient amount, say 10 ml. If the 100-ml sample is concentrated to 10 ml, it has been concentrated by a factor of ten. Acidify the solution with dilute nitric acid, if insoluble materials have separated. This solution can be analyzed as in Parts 1-6. Use solution without added nitric acid to test for nitrate ion. Note that tests for many ions can be done directly on the unknown solution without going through the hydrogen sulfide separation. These are Ca^{2+} and Mg^{2+} (with ammonium carbonate), Fe^{3+} (with KSCN), Cu^{2+} (with NH_3), Ag^+ (with HCl), and Pb^{2+} (with HCl), as well as Cl^-, SO_4^{2-}, CO_3^{2-}, and NO_3^-. Also, magnesium salts are common constituents of drinking water, and Mg^{2+} is not one of the cations of the scheme in Parts 1-6. In the analytical scheme, Mg^{2+} forms a carbonate, as does Ca^{2+} (see Table 26-1), which is also an alkaline earth metal. Consequently, this scheme does not distinguish between these two ions. The approximate concentration of various ions in the water can be estimated by comparing the amount of precipitate or color intensity with those obtained with standard solutions that are mostly $0.02M$.

SUPPLEMENTARY READING

Dickerson, Gray, and Haight (Chapters 16 and 17)

Masterton and Slowinski (Chapter 16, Section 4, and Chapter 18, Section 5)

Sienko and Plane (Chapter 12, Sections 8 and 9)

Mahan (Chapter 6, Section 5)

Brescia, *et al.* (Chapter 20)

Brown (Chapter 15, Section 11)

Hammond, Osteryoung, Crawford, and Gray (Chapter 11)

QUESTIONS

1. Which of the cations in this experiment form insoluble chlorides?
2. How can lead ion be separated from silver ion?
3. How can Fe^{2+} and Ni^{2+} be separated from Cu^{2+}, Hg^{2+}, and As^{3+}? On what difference in property does this separation depend? (*Note*: FeS and NiS subsequently are separated from one another by adding dilute HCl. FeS dissolves readily, but NiS does not, although neither NiS nor FeS precipitated in acidic solution of H_2S. The reason for this paradoxical phenomenon is that the NiS solid changes structure on standing or heating to a form that dissolves slowly in HCl.)
4. How is Al^{3+} separated from Fe^{3+}? Explain with an equation. This same property is used to purify bauxite (Al_2O_3) from impurities of iron oxide (Fe_2O_3).
5. What happens when Fe^{2+} ion in solution is shaken with air?
6. How is arsenic separated from Cu and Hg?
7. How can you separate copper ions from mercuric ions?
8. An unknown solution is colorless. Which cations must be absent?
9. What cations cannot be present in your unknown solution if the solution contains a high concentration of chloride ion?

PROBLEMS

1. Copper sulfide (CuS) is much less soluble than iron sulfide (FeS), as indicated by the solubility products in Table 26-1. Calculate the hydronium ion concentration needed in a $0.1M$ solution of hydrogen sulfide to precipitate CuS but not FeS. The initial concentrations of Cu^{2+} and Fe^{2+} are $0.02M$ each.

2. A solution that may contain any or all of the cations in this assignment (Ag^+, Pb^{2+}, Al^{3+}, Fe^{3+}, Hg^{2+}, As^{3+}, Ni^{2+}, Cu^{2+}, NH_4^+, Ca^{2+}) is treated with HCl and no precipitate results. The solution is then treated with H_2S, in acidic solution, and a precipitate results. A portion of this precipitate dissolves in basic H_2S solution, and this solution deposits a tan precipitate on acidification. The remainder of the precipitate dissolves in hot nitric acid. When this solution is treated with excess ammonia, a blue solution results.

The decantate from the first H_2S treatment is saturated with H_2S in slightly basic solution. A precipitate forms, which dissolves completely in aqueous HCl after separation. When the solution is made basic with excess NaOH, a brown precipitate results. When this precipitate is removed, and the solution is buffered with NH_4Cl, a gelatinous white precipitate results. List the cations present. List the cations proved not to be present, and also list the cations that might or might not be present.

3. How could you distinguish between three samples of silver or silver alloy, all of which look alike, by chemical methods. One is a silver-copper alloy, and one is a silver-lead alloy. The third sample is pure silver. Also, describe how you would dissolve them for analysis.

4. An industrial company producing semiconductor devices had a problem with the leads (connecting wires) of diodes developing a black scale while in storage on the East coast. In fact, a large customer was about to cancel the contract because the leads had to be cleaned before they could be soldered. The leads are copper wire plated with nickel. The diode leads are held between strips of adhesive tape so that the diodes can be stored on rolls. A chemist finally attempted to solve the problem. He scraped off some of the scale and dissolved it in hot concentrated nitric acid. Addition of ammonia to the solution resulted in a deep blue color. Addition of barium chloride solution produced a white precipitate. What is the black scale? (Incidentally, the tape had a sulfur adhesive, and heat and moisture catalyzed the formation of the scale. The problem was avoided by using a different tape.)

EXPERIMENT 27

Equipment and supplies

One potentiometer (see procedure, cost $45) or vacuum tube voltmeter, capable of reading to 0.01 volt

Two wires with alligator clips

Two 100-ml beakers

One zinc electrode

Two copper electrodes (about 3.5 × 1 × 1/16 inch cut from bar stock)

One glass U-tube of 6-8 mm glass

One 10-ml graduated cylinder

One 100-ml graduated cylinder

One medicine dropper

100 ml 0.10M Zinc sulfate

200 ml 0.10M Copper sulfate

10 ml 0.5M Potassium chloride

20 ml 6M Hydrochloric acid

20 ml 6M Nitric acid

100 ml 1.0M Potassium hydroxide

100 ml 1.0M Ammonia

Cotton

Time requirement

Two hours

ELECTROCHEMICAL CELLS

INTRODUCTION

If a copper strip is placed in a solution of copper ions, one of the following reactions may occur

$$Cu^{2+} + 2e^- \rightarrow Cu$$

$$Cu \rightarrow Cu^{2+} + 2e^-$$

The electrical potential that would be developed by these reactions prevents their continuation. These reactions are called *half-reactions* or half-cell reactions. There is no direct way to measure the electrical potential (electromotive force, emf) of a half-cell reaction. Similarly, a zinc strip in a solution of zinc ions has the possible reactions

$$Zn^{2+} + 2e^- \rightarrow Zn$$

$$Zn \rightarrow Zn^{2+} + 2e^-$$

but these also are prevented from occurring by the electrical potential that would build up. If the metal electrodes (copper and zinc) in the two solutions are connected by a wire, and if the solutions are electrically connected by perhaps a porous membrane or a bridge that minimizes mixing of solutions, a flow of electrons will move from one electrode, where the reaction is

$$M_1 \rightarrow M_1^{n+} + ne^-$$

to the other electrode, where the reaction is

$$M_2^{n+} + ne^- \rightarrow M_2$$

In this case, the zinc metal goes into solution as zinc ions and the copper ions are plated out. The overall cell reaction is

$$Zn + Cu^{2+} \rightarrow Zn^{2+} + Cu$$

The electromotive force for such a cell, which is written

$$Zn \mid ZnSO_4 \parallel CuSO_4 \mid Cu$$

can be measured.

By convention, all half-cell emf's are compared to the emf of the standard hydrogen electrode. The standard hydrogen electrode is defined as a platinum electrode covered with platinum black that is in contact with hydrogen gas at 1 atmosphere pressure and a 1 molal solution of hydronium ions (actually, it is defined for unit activity). The hydrogen electrode half-cell reaction is

$$H_2 \rightleftarrows 2H^+ + 2e^-$$

The emf of a half-cell, with respect to the standard hydrogen electrode, is called the *oxidation potential*. Standard oxidation potentials, $\mathcal{E}^0$, are for 1 molal solutions. Consequently, the difference between the oxidation potentials of two half-cells is the emf they would develop if connected together as a cell. The emf for the Zn-Cu cell described would be

$$\mathcal{E} = \mathcal{E}^0(Zn) - \mathcal{E}^0(Cu) = 0.77 - (-0.34) = 1.10 \text{ V}$$

if the solutions are 1.0 molal.

If a cell reaction can be written

$$aA + bB \rightarrow cC + dD$$

then the emf of the cell can be expressed in the form of the following equation, developed by Nernst[53]:

$$\mathcal{E} = \mathcal{E}^0 - \frac{2.3RT}{n\mathcal{F}} \log \frac{[C]^c[D]^d}{[A]^a[B]^b}$$

If all the concentrations are $1m$, then the logarithmic term becomes zero and

$$\mathcal{E} = \mathcal{E}^0$$

which is the reason for choosing 1-molal concentration as the standard condition. If the system is at equilibrium, then

$$\frac{[C]^c[D]^d}{[A]^a[B]^b} = K_{eq}$$

and the emf developed by such a cell at equilibrium must be zero. Therefore,

$$0 = \mathcal{E}^0 - \frac{2.3RT}{n\mathcal{F}} \log K_{eq}$$

or

$$\log K_{eq} = \frac{n\mathcal{F}\mathcal{E}^0}{2.3RT}$$

In these equations, $\mathcal{F}$ is the Faraday, n is the number of electrons transferred in the oxidation-reduction step, and R is the gas constant in units of electrical work.

MEASURING EMF

To measure the emf of a galvanic cell, a sensitive meter is needed, but it is important that the meter not draw a significant amount of current. If the current passed by the cell to be measured is large, the cell will become polarized, and the emf will decrease. Consequently, to measure minute currents in a cell of reasonable size, we shall use a device that applies a known emf in opposition to the unknown cell emf. This type of instrument was developed first by Poggendorf.[54] The condition when the two emf's are just equal and opposite (null point) will be detected with a very sensitive galvanometer. A 10-turn linear helipot will be used as the voltage divider, R. The circuit diagram for this instrument[55] will be found in Figure 27-1. A photograph of the instrument connected to a cell with a U-tube is included as Figure 27-2. If you have some knowledge of electricity (Ohm's law is sufficient), it will be enlightening to study these circuits. The actual operation of the potentiometer is much simpler than the circuit diagram perhaps would suggest. However, do not attempt to use the instrument until the instructor has "checked you out" on it.

A vacuum tube voltmeter can be used to measure the voltages in this experiment. However, few freshman students will be able to understand its operational principles, and for this reason the Poggendorf potentiometer is used here. The potentiometer is also more accurate.

The accuracy of this potentiometer is limited by the accuracy of the standard cell, which is a 1.34-V mercury cell, and by the tolerance of the resistors in the voltage divider network. The mercury cell has an accuracy of about ±1/2%, and the resistors in the voltage divider have 1% tolerance. However, the 10-turn linear helipot is linear to ±0.25%. Consequently, although measurements are only accurate to about 1% absolute, measurements (compared to one another) have about the accuracy of ±0.25% of the voltage range.

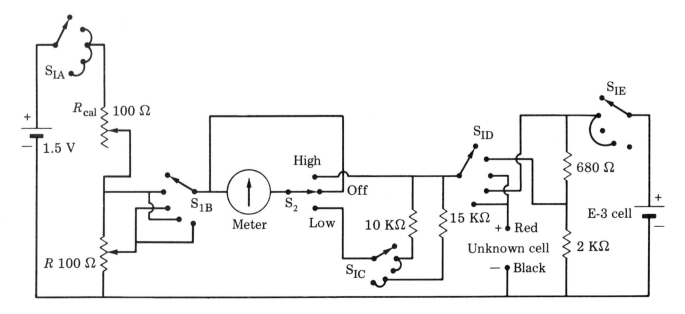

Figure 27-1. Circuit diagram of potentiometer.

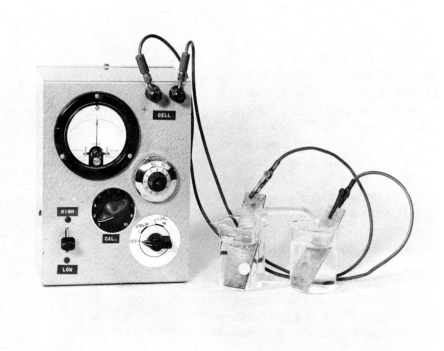

Figure 27-2. Potentiometer connected to a galvanic cell.

SIMPLIFIED EXPLANATION OF THE POTENTIOMETER

When the potentiometer is in the calibrating position (when the selector switch is on calibrate, C) the circuit can be represented simply as in Figure 27-3. Notice that the standard cell is connected in opposition to the driving cell. To calibrate the potentiometer, the variable resistance, R_{cal}, is adjusted so that the galvanometer shows no deflection. When the galvanometer shows no deflection, no current is flowing through it, which means that the voltage between Points A and B across the resistance R is equal exactly to that of the standard cell

$$V_{AB} = V_{STD}$$

Current is not flowing through the circuit loop containing the galvanometer, but current is flowing in the circuit loop containing the driving cell, R_{cal}, and Points A and B. From Ohm's law the voltage between A and B must equal the resistance between the Points A and B, multiplied by the current flowing

$$V_{STD} = V_{AB} = IR_{AB}$$

Now the selector switch is changed to the measuring position, M, in which the unknown cell is connected in opposition to the driving cell. The circuit is represented simply as in Figure 27-4. R_{cal} is left as it is. The variable resistance R, the 10-turn helipot, now is adjusted so that the galvanometer shows no deflection. Then no current flows through the circuit loop containing the galvanometer, therefore the

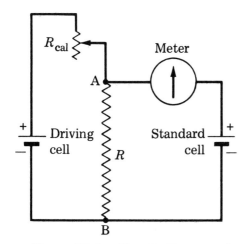

Figure 27-3. Circuit diagram of potentiometer in calibrating position.

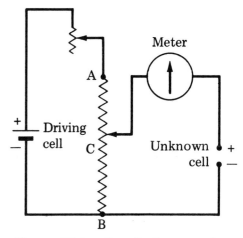

Figure 27-4. Circuit diagram of potentiometer in measuring position.

voltage of the unknown cell must be equal to the voltage between Points C and B of the variable resistance R

$$V_X = V_{CB}$$

The voltage V_{CB} must be equal to the resistance between Points C and B multiplied by the current flowing in the circuit loop connecting the driving cell, R_{cal}, and Points A, C, and B

$$V_X = V_{CB} = IR_{CB}$$
$$V_X = IR_{CB}$$

Combining this relationship with the calibrating condition, $V_{STD} = IR_{AB}$, gives

$$\frac{V_X}{V_{STD}} = \frac{R_{CB}}{R_{AB}}$$

$$V_X = \frac{R_{CB}}{R_{AB}} V_{STD}$$

The fraction R_{CB}/R_{AB} is a fraction of the total resistance of the helipot, R. The helipot dial reads from 0 to 1000, and is arranged so that the number read from the dial is $(R_{CB}/R_{AB})1000$. Consequently, if the dial reads 810, R_{CB}/R_{AB} is 810/1000. The unknown voltage can be calculated $V_X = 0.810\ V_{STD} = 0.810(1.34) = 1.09$ V.

PROCEDURE

Part 1: The Daniell cell[56]

Place $0.1M$ $ZnSO_4$ solution in a 100-ml beaker, and $0.1M$ $CuSO_4$ solution in another beaker of the same size. The liquid levels should be the same. In the zinc solution place a clean strip of zinc, and in the copper solution, a clean strip of copper. The zinc strip may be cleaned by placing it in a beaker of about $2M$ HCl (made by diluting one volume of $6M$ HCl with two volumes of water). The copper strip can be cleaned by placing it in a beaker of $2M$ HNO_3 (made by diluting one volume of $6M$ HNO_3 with two volumes of water). Rinse the electrodes with distilled water thoroughly before using them.

Attach the wires to the electrodes, connecting them to the potentiometer. The black terminal is negative and the red terminal is positive. Connect the wire from the zinc electrode to the black terminal on the potentiometer, and the copper to the red terminal. The cells are connected with a U-tube filled with $0.5M$ potassium chloride and closed with cotton stuffed in the ends (see Figure 27-5). Electrons will flow from the zinc electrode to the copper electrode, because of the reactions occurring in the half-cells

$$Zn \rightarrow Zn^{2+} + 2e^-$$

$$Cu^{2+} + 2e^- \rightarrow Cu$$

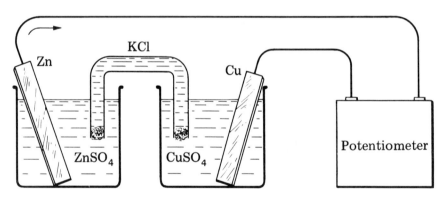

Figure 27-5. The Daniell cell.

The overall reaction for the cell is

$$Cu^{2+} + Zn \rightarrow Cu + Zn^{2+}$$

The cell may be represented as

$$Zn\,|\,0.1M\ ZnSO_4\,\vdots\,0.5M\ KCl\,\vdots\,0.1M\ CuSO_4\,|\,Cu$$

Measure the voltage of this cell as described below, using the 1.34-V scale. Also, measure the voltage of a Daniell cell with $0.01M$ $ZnSO_4$ and $0.1M$ $CuSO_4$, and with $0.1M$ $ZnSO_4$ and $0.01M$ $CuSO_4$.

Use of the potentiometer

Calibration. Connect the unknown cell to the indicated terminals, taking care to observe the correct polarity. Turn the selector switch (S_1) from the "off" position to the "calibrate" Position, C. Push the lever switch (S_2) from off to the low sensitivity position. Adjust R_{cal}, the variable resistance for standardizing the voltage of the potentiometer versus the standard cell, so that the galvanometer shows no deflection, that is, the dial reads 0. When the R_{cal} has been adjusted so that the galvanometer reads nearly zero, switch the lever switch (S_2) to high sensitivity, and further adjust R_{cal} so that the galvanometer reads zero. Release the lever switch. It will spring back to "off." The potentiometer now is calibrated. Do not change R_{cal}, or the calibration is spoiled. Immediately go on to the next section.

Measuring the emf of the unknown cell. Turn the selector switch (S_1) to "measure," M. Push the lever switch (S_2) to low sensitivity and adjust the 10-turn helipot (R) until the galvanometer deflection is near zero. Then push the lever switch (S_2) to high sensitivity and adjust the helipot (R) further, until the galvanometer reads zero deflection. This means no current is passing through it. When this is done, let the lever switch (S_2) spring back to off. Turn the selector switch back to "calibrate," and check the calibration as described previously. If the calibration is still very good, record the dial reading of the helipot, R. It is necessary to check calibration before and after a measurement, because the emf of the driving cell may change due to polarization or depolarization. If the calibration has changed, recalibrate, measure the unknown cell, and check the calibration. The reading of the 10-turn dial of R was recorded. The maximum reading on the dial is 1000, which corresponds to 1.34 V, if the selector switch is on the 1.34-V scale (and to 1.00 V, if the selector switch is on the 1.00-V scale). To find the number of volts, multiply the number you read from the dial by 1.34 V and divide by 1000.

Part 2: Solubility product of cupric hydroxide

If an electrochemical cell can be built in which the half-cell reactions are

$$Cu^{2+} + 2e^- \rightleftarrows Cu$$

$$Cu + 2OH^- \rightleftarrows Cu(OH)_2(s) + 2e^-$$

the overall cell reaction is the reverse of the solubility product of cupric hydroxide

$$Cu^{2+} + 2OH^- \rightleftarrows Cu(OH)_2(s)$$

Such a cell can be written as

$$Cu \mid Cu^{2+} \mid KCl \mid OH^- \mid Cu(OH)_2 \mid Cu$$

From the Nernst equation,

$$\mathscr{E} = \mathscr{E}^0 - \frac{2.3RT}{2\mathscr{F}} \log \frac{1}{[Cu^{2+}][OH^-]^2}$$

$$\mathscr{E} = \mathscr{E}^0 + \frac{2.3RT}{2\mathscr{F}} \log [Cu^{2+}][OH^-]^2$$

$$\mathscr{E}^0 = \mathscr{E} - \frac{2.3RT}{2\mathscr{F}} \log [Cu^{2+}][OH^-]^2$$

Consequently, if we can measure the emf, $\mathscr{E}$, of such a cell, we can calculate $\mathscr{E}^0$, if we know the concentration of cupric ion in one half-cell and the concentration of hydroxide ion in the other half-cell.

In the introductory discussion, we have shown from the Nernst equation that the following relationship exists

$$\log K = \frac{n\mathcal{F}\mathcal{E}^0}{2.3RT}$$

So if $\mathcal{E}^0$ is obtained, it can be converted into a value of the solubility product. The factor $2.3RT/\mathcal{F}$ is equal to 0.059 V at 25°C.

Construct a cell of two beakers. In a 100- or 150-ml beaker, place 0.10M CuSO$_4$ solution and a clean copper strip. The copper strip can be cleaned in dilute nitric acid, but it must be washed well with distilled water. Until it is put into the copper sulfate solution, it should be kept in dilute HCl, but rinsed with distilled water before use. In a clean beaker, add the same height of 0.10M potassium hydroxide solution. Add a copper electrode to this KOH solution. Connect wires to the electrodes and the appropriate terminals of the potentiometer. The KOH side is the negative electrode, because copper goes into the oxidized form, Cu^{2+}, because of the low concentration of Cu^{2+} in the hydroxide solution. Now add the salt bridge, which is a U-tube filled with KCl and closed off by wads of cotton in each end. The cotton wads should be soaked. Measure the emf of the cell as soon as possible. The cell is polarized very easily. To get around this difficulty, clean the electrodes again, rinse, and return them to the solutions. Remeasure the voltage, starting from the last position of the helipot dial. The new null point will be somewhat higher if the cell was polarized. Measure the emf of the cell, which uses 0.10M cupric sulfate and 1.0M potassium hydroxide. Calculate the solubility product, K_{sp}, using the equations above. Literature values range from 10^{-14} to 10^{-20}, the latter being considered the best value. Taking the standard $\mathcal{E}^0$ of the Cu-Cu^{2+} electrode to be -0.34 V, evaluate $\mathcal{E}^0$ for the Cu-Cu(OH)$_2$-OH$^-$ electrode.

Part 3: Formation constant of tetraamminecopper(II) cation

Similarly, if we can construct a cell having the half-reactions

$$Cu^{2+} + 2e^- \rightleftarrows Cu$$

$$Cu + 4NH_3 \rightleftarrows Cu(NH_3)_4^{2+} + 2e^-$$

the overall cell reaction is the formation equilibrium for tetraamminecopper(II) cation

$$Cu^{2+} + 4NH_3 \rightleftarrows Cu(NH_3)_4^{2+}$$

Such a cell can be written

$$Cu \mid Cu^{2+} \vdots KCl \vdots NH_3, Cu(NH_3)_4^{2+} \mid Cu$$

The Nernst equation can be written

$$\mathcal{E} = \mathcal{E}^0 - \frac{2.3RT}{2\mathcal{F}} \log \frac{[Cu(NH_3)_4^{2+}]}{[Cu^{2+}][NH_3]^4}$$

$$\mathscr{E}^0 = \mathscr{E} + \frac{2.3RT}{2\mathscr{F}} \log \frac{[\mathrm{Cu(NH_3)_4^{2+}}]}{[\mathrm{Cu^{2+}}][\mathrm{NH_3}]^4}.$$

and the formation constant, K_f, is related by

$$\frac{2\mathscr{F}\mathscr{E}^0}{2.3RT} = \log K_f$$

Clean the copper strips in $6M$ hydrochloric acid. To a 100-ml beaker, add 50 ml of $1M$ aqueous ammonia and 0.25 ml (this small volume may be measured using a calibrated medicine dropper, probably about 5 drops) of $0.10M$ cupric sulfate solution. Mix well and add a rinsed copper electrode. To another 100-ml beaker, add 50 ml of $0.1M$ cupric sulfate and another rinsed copper electrode. Connect the electrodes to the potentiometer. Which electrode is negative? Make the electrical connection by adding the salt bridge. Determine the voltage of the cell. Using the calibrated medicine dropper, add 2.25 ml (probably about 45 drops) more of $0.1M$ cupric sulfate to the beaker containing cupric ammine complex, mix well, and determine the voltage of the cell. Assuming that the reaction of the 0.25 ml of $0.10M$ cupric sulfate solution and $1M$ aqueous ammonia goes to completion

$$\mathrm{Cu^{2+}} + 4\mathrm{NH_3} \rightarrow \mathrm{Cu(NH_3)_4^{2+}}$$

calculate the concentration of cupric ammine complex in the first cell. Using this value for $[\mathrm{Cu(NH_3)_4^{2+}}]$, $[\mathrm{Cu^{2+}}] = 0.10M$, and $[\mathrm{NH_3}] = 1$, calculate $\mathscr{E}^0$ from the preceding equation, and then calculate K_f. Do the same for the second cell.

SUPPLEMENTARY READING

Dickerson, Gray, and Haight (Chapter 18)

Masterton and Slowinski (Chapter 20, and Chapter 21, Section 5)

Sienko and Plane (Chapter 13)

Mahan (Chapter 7, Sections 4, 5, and 7)

Brescia, *et al.* (Chapter 16, and Chapter 19, Sections 12 and 19)

Brown (Chapter 16)

Pauling (Chapter 11, Section 6, and Chapter 18, Sections 10 and 11)

Hammond, Osteryoung, Crawford, and Gray (Chapters 10 and 11)

QUESTIONS

1. Describe the reactions taking place at the electrodes, and any other processes necessarily occurring in the Daniell cell while it is operating. Note that the electrode reactions produce and consume ions at the electrodes.

2. How should the voltage change in the Daniell cell if the zinc solution is made more dilute (and the concentration of the copper solution is kept the same)?

3. How will a large air bubble in the salt bridge affect the operation of the cell? Can you suggest a reason?

4. Design a cell whose voltage, when measured, can be used to calculate the solubility product of silver chloride

$$AgCl(s) \rightleftarrows Ag^+ + Cl^-$$

5. Design a cell whose voltage, when measured, can be used to calculate the formation constant of the diamminesilver cation

$$Ag^+ + 2NH_3 \rightleftarrows Ag(NH_3)_2^+$$

PROBLEMS

1. Gold does not dissolve in nitric acid. However, in the mixture of nitric acid and hydrochloric acid, called aqua regia, gold does dissolve, forming the tetrachloroaurate(III) complex ion and nitrogen oxides. Explain the reasons for this behavior from the following oxidation potentials:

$$Au \rightarrow Au^{3+} + 3e^-; \mathcal{E}^0 = -1.42 \text{ V}$$

$$Au + 4Cl^- \rightarrow AuCl_4^- + 3e^-; \mathcal{E}^0 = -1.00 \text{ V}$$

Also, calculate the formation constant for the gold chloride complex ion.

2. What purpose does the salt bridge serve in the electrochemical cell? Discuss alternative possibilities.

3. Calculate the solubility product of silver chloride, an extremely insoluble salt, from the following oxidation potentials:

$$Ag \rightarrow Ag^+ + e^-; \mathcal{E}^0 = -0.7996 \text{ V}$$

$$Ag + Cl^- \rightarrow AgCl + e^-; \mathcal{E}^0 = -0.2221 \text{ V}$$

4. Devise an electrochemical cell or cells that can be utilized to measure the solubility product of ferric hydroxide.

5. Devise an electrochemical cell that can be used to measure hydronium ion concentration, or the logarithm of hydronium ion concentration—the pH.

EXPERIMENT 28

Equipment and supplies

One gas burette (constructed of two 50-ml burettes)

4 feet plastic tubing (see procedure)

One 250-ml Erlenmeyer flask

Pieces of glass tubing or boiling stones

One 100-ml graduated cylinder

Clock with second hand

Barometer

Thermometer

250 ml 0.2M Hydrogen peroxide

20 ml 6M Hydrochloric acid

40 ml 0.005M Ferric chloride

Glycerol

Time requirement

Two hours

CHEMICAL KINETICS; DECOMPOSITION OF HYDROGEN PEROXIDE

The purpose of this experiment is to determine the effects of concentration, temperature, and catalysts on the rate of decomposition of hydrogen peroxide.

HISTORICAL NOTE

The fact that reactions take time to occur has been known for a long time, in smelting and brewing, for example. Furthermore, it was known that the fermentation in making beer proceeds faster at higher temperature; that is, more sugar is converted into alcohol and carbon dioxide in a given length of time at higher temperature than at lower temperature.

In 1777, C. F. Wenzel found that if a certain concentration of acid dissolved a gram of zinc in one hour, a solution of acid half as strong required two hours, provided the metal surface and the temperature were the same. In more general terms, the rate of reaction depends on the concentration of the reactants.

In 1840, L. Wilhelmy[39] reported his studies on the inversion of sucrose (ordinary cane sugar); he followed the reaction rate with a polarimeter. He found that the rate of decrease of sucrose concentration was proportional to the concentration of sucrose, that is,

$$-\frac{dA}{dt} = k_1 A$$

where A is the concentration of sucrose, and k_1 is the reaction constant. The expression dA/dt in the equation means the instantaneous rate of change of A in a very small time interval, and approximates what is usually meant by $\Delta A/\Delta t$ for larger time intervals.

Wilhelmy integrated the above differential equation to

$$\ln A = -kt + \text{constant}$$

from which the value of the reaction rate constant could be evaluated. It should be noted that this equation is of the form $y = ax + b$, which is the equation of a straight line. Therefore, if the logarithms of the concentration of reactant at various times are plotted versus time, the points should make a straight line. The slope of the line will be $-k$ if natural logarithms are used, and $-k/2.3$ if logarithms to the base 10 are used.

Berthelot and St. Gilles[40] investigated the equilibrium between ethyl acetate and water, and between acetic acid and ethanol. Shortly afterward, Guldberg and Waage[41] interpreted these data, and others, in terms of the theoretical reactions

$$A + B \rightarrow C + D$$

and

$$C + D \rightarrow A + B$$

where

$$\text{rate forward} = k_f[A][B]$$

$$\text{rate backward} = k_r[C][D]$$

At equilibrium, the rate of reaction in the forward direction is equal to the rate of reaction in the backward direction, and therefore

$$k_f[A][B] = k_r[C][D]$$

Rearranging,

$$K = \frac{k_f}{k_r} = \frac{[C][D]}{[A][B]}$$

They generalized the equilibrium as

$$aA + bB \rightleftarrows cC + dD$$

$$K = \frac{[C]^c[D]^d}{[A]^a[B]^b}$$

In 1884 J. H. van't Hoff published *Étude de Dynamique Chimique*,[57] in which methods were described for determining reaction order, and in which thermodynamics was applied to chemical equilibrium. He derived the relationship between the equilibrium constant and absolute temperature

$$\frac{d \ln K_c}{dT} = \frac{q}{2T^2}$$

where q is the energy difference per mole between reactants and products. Integration gives

$$\ln K_c = -\frac{q}{2T} + \text{constant}$$

for which the exponential form would be

$$K_c = e^{-q/2T}(e^C)$$

where e^C is the constant. It might seem that since the equilibrium constant is the ratio of forward and reverse rate constants

$$K = \frac{k_f}{k_r}$$

one could derive the temperature dependence of the rate constant from the above equations, as follows:

$$\frac{d \ln k_f}{dT} - \frac{d \ln k_r}{dT} = \frac{q}{2T^2}$$

Van't Hoff pointed out that the dependence of k on T then must be of the form

$$\frac{d \ln k}{dT} = \frac{A}{T^2} + B$$

where B might be a function of temperature. Therefore, the desired relationship, he said, cannot be obtained in this way.

At about this time, it was pointed out that the rate constant of a chemical reaction, ρ, changed greatly for a small change in temperature.[58] A relationship

$$\rho = a\alpha^\Theta$$

was suggested, where Θ is temperature, and a and α are constants. Other relationships had been tried, such as

$$\rho = a\Theta^n$$

where n is an arbitrary number.

Svante Arrhenius showed[59] that the rate constants of many reactions in the literature (including the inversion of sucrose) were described very well by the following equation:

$$\rho_1 = \rho_0 \exp \frac{A(T_1 - T_0)}{T_0 T_1}$$

Arrhenius was able to derive this relationship by postulating that not all molecules react, but that only "active" molecules react. Energy (q) is required to make an active molecule, a, from an inactive molecule, i.

$$i + q \rightleftarrows a$$

He expressed the equilibrium between active molecules and inactive molecules in terms of k, an equilibrium constant, and the concentration of active molecules $[a]$ and inactive molecules $[i]$.

$$k = \frac{[a]}{[i]}$$

which should depend on temperature according to Van't Hoff's equation

$$k_1 = k_0 \exp \frac{q(T_1 - T_0)}{2T_1 T_0}$$

Since k is essentially the rate constant (or proportional to it), we can rewrite the relationship in modern form

$$k = A \exp \frac{-E_a}{RT}$$

where k is now the rate constant; A, a constant; E_a, the energy of activation; R, the gas constant, 1.98 cal mole^{-1} deg^{-1}; and T, the absolute temperature. The logarithmic form is

$$\ln k = \ln A - \frac{E_a}{RT}$$

which is a linear equation. If the logarithms of rate constants at various temperatures are plotted against T^{-1}, the slope of the line is $-E_a/R$.

INTRODUCTION

Hydrogen peroxide decomposes to form oxygen gas and water, according to the equation

$$2H_2O_2 \rightarrow O_2 + 2H_2O$$

This reaction is slow, but is catalyzed by ferric chloride. Catalysts include other transition metal ions and potassium iodide. It has been found that the reaction velocity, or rate, depends on the concentration of peroxide, and on the concentration of ferric ion

$$\text{rate} = k_2[H_2O_2][Fe^{3+}]$$

where k_2 is the second-order rate constant. Because the ferric ion concentration is constant during the reaction, the rate equation may be written as a first-order rate law, in which k_1 contains k_2 and the ferric ion concentration

$$\text{rate} = \frac{-\Delta[H_2O_2]}{\Delta t} = k_1[H_2O_2]$$

This rate law states that the amount of peroxide decomposed in a unit of time is proportional to the concentration of peroxide at that instant. If this equation is integrated with respect to time, the following integrated rate equation is obtained

$$\ln[H_2O_2] = k_1 t + C$$

or

$$2.3 \log[H_2O_2] = k_1 t + C$$

where C is a constant of integration, and the concentration of peroxide is that at time t.

PROCEDURE

This reaction will be followed by measuring the volume of the evolved oxygen with a gas burette. The gas burette is constructed of two 50-ml burettes connected at the bottom ends with a 75-cm piece of plastic tubing. Another piece (1 meter) of plastic tubing fitted with rubber stoppers connects the top of one burette with the reaction flask. At the start of the reaction, the burette should be filled with water so that the water level is at the 0-ml mark (full) and at the 50-ml mark (empty) in the other burette, which serves as a leveling reservoir. Figure 28-1 shows the apparatus at the beginning of a run. The burettes may be held conveniently in a double burette clamp between runs. The burette connected to the reaction flask is left clamped when the reaction is in progress, and the reservoir burette is held by hand.

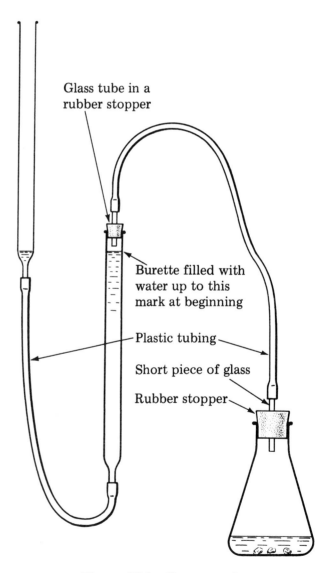

Figure 28-1. Gas-measuring apparatus at the beginning of an experiment.

Put 50 ml of 0.2M H_2O_2 in a 125- or 250-ml Erlenmeyer flask that contains pieces of glass or Carborundum boiling stones. Record the volume, concentration, and temperature of the solution. Connect the flask to the gas burette. With one hand swirl the flask at a regular rate. Use your other hand to keep the water in the two burettes at the same level so that the pressure of the oxygen is atmospheric pressure. To obtain the best data, it is desirable to swirl the flask continually to prevent supersaturation of the solution with oxygen. Since you do not have three hands, set the reaction flask down for a few seconds when recording the burette reading (±0.1 ml) every 60 seconds. Record the results of a 10-minute reaction duration in tabular form. The total volume of oxygen evolved in this period should be one ml or less. If the flask contained slight traces of ferric chloride more than one ml of oxygen will be produced. If that is the case, rinse the flask, stopper, and stones carefully with dilute hydrochloric acid and then with distilled water several times. Then repeat the experiment.

Rinse the reaction flask and add 40 ml of 0.2M H_2O_2 and 10 ml of 0.005M ferric chloride. Record the temperature of the solution. Connect the flask to the gas burette and begin agitation of the solution. Record the initial reading of the burette and the reading every minute. Continue the reaction until about 50 ml of gas is evolved.

Rinse the reaction flask. Fill a plastic pan with water at 35°C-40°C and record the temperature. Add 40 ml of 0.2M hydrogen peroxide and 10 ml of 0.005M ferric chloride to the flask. Connect the flask to the gas burette and swirl it in the pan of warm water. Tabulate the results of the reaction at this temperature. Record the final temperature of the bath. This temperature should not have fallen more than three degrees.

Do another experiment with a different concentration of peroxide or ferric chloride. Other temperatures may be tried. The catalytic effect of avocado peel on the decomposition of hydrogen peroxide also is interesting.

Note: Approximately 0.2M hydrogen peroxide can be made by adding one liter of 30% hydrogen peroxide to 43 liters of water. If the dilute 0.2M hydrogen peroxide is not freshly prepared, it should be standardized roughly by titration of 10 ml with 0.04M potassium permanganate solution in the presence of dilute sulfuric acid. See Experiment 14 for further details. For the purpose of this experiment, the concentration to two significant figures is sufficient. The balanced equation for the titration is

$$2MnO_4^- + 5H_2O_2 + 6H^+ \rightarrow 5O_2 + 2Mn^{2+} + 8H_2O$$

Store hydrogen peroxide only in bottles with blowout caps.

CALCULATIONS

Arrange your data in a table having eight columns. In Table 28-1, sample data are given for part of a reaction. List the elapsed time and the burette readings in the first two columns. In the third column, enter

Table 28-1.

Time elapsed, minutes	Burette reading	Gas produced, ml	O$_2$ produced, mmoles	H$_2$O$_2$ consumed, mmoles	H$_2$O$_2$ left in solution, mmoles	[H$_2$O$_2$] left	log[H$_2$O$_2$]
—	1.0	0	0	0	7.50	0.125	−0.903
0	7.3	6.3	0.248	0.496	7.00	0.117	−0.932
1	9.9	8.9	0.350	0.700	6.80	0.113	−0.947
2	12.3	11.3	0.444	0.888	6.61	0.110	−0.959

the volume of gas produced, as calculated from the second column. Calculate the number of millimoles of oxygen in 1.00 ml of gas, taking into account the temperature, atmospheric pressure, and vapor pressure of water

$$\left(\frac{1000 \text{ mmoles}}{\text{mole}}\right) \times \left(\frac{0.001 \text{ liter}}{22.4 \text{ liters mole}^{-1}}\right) \times \left(\frac{P - P_{H_2O}}{760 \text{ mm atm}^{-1}}\right) \times \left(\frac{273}{T}\right) = \frac{\text{mmoles O}_2}{\text{ml gas}}$$

Now convert the volumes of gas produced (the third column) to millimoles of oxygen produced (fourth column). The fifth column is the number of millimoles of hydrogen peroxide consumed, which is twice the number of millimoles of oxygen produced. In the run given as an example, there was 7.50 millimoles of hydrogen peroxide at the beginning of the reaction, and the volume of the reaction solution was 60 ml. The sixth column is the number of millimoles of peroxide left at each recorded time. Columns seven and eight are for the concentration and logarithm of concentration of peroxide at each time. In order to keep this calculation from taking the whole night, use a slide rule. Also, it is faster to do the same operation for the entire column of points before going on to the next column. Plot your data, with the concentrations along the vertical axis and the time along the horizontal axis. The slope of this plot is the rate of reaction. Now plot the logarithm (base 10) of peroxide concentration against time. The slope of this line is −(1/2.3)k_1.

For the other runs that you make, simplify the work by calculating only about 10 or 15 points, evenly spaced over the reaction time. Then plot only the logarithm of concentration against time to obtain the rate constant. Plot the logarithms (base 10) of your rate constants for two temperatures (or more) versus the reciprocal of the absolute temperatures. (For these rate constants, k_1, the run must have the same concentration of catalyst, or the constant must be divided by the catalyst concentration.) The slope of this plot is $E_a/2.3R$. Calculate the activation energy from your plot.

SUPPLEMENTARY READING

Dickerson, Gray, and Haight (Chapter 19)

Masterton and Slowinski (Chapter 14)

Sienko and Plane (Chapter 10)

Mahan (Chapter 9, Sections 1 and 5)

Brescia, *et al.* (Chapter 23)

Brown (Chapter 13)

Pauling (Chapter 18)

Hammond, Osteryoung, Crawford, and Gray (Chapter 13)

QUESTIONS

1. How does the rate of reaction (slope of concentration versus time) change with time?
2. Why?
3. How does the rate constant, k_1, (slope of logarithm of concentration versus time) change with time?
4. How does the rate constant change with temperature?
5. From your measurements, what arguments can be made in favor of the first-order rate law and against the second-order rate law, for this reaction?

APPENDIXES

APPENDIX 1/ REFERENCES

1. W. Prout, *Ann. Philosophy* **6**, 321 (1815); **7**, 111 (1816).
2. J. J. Berzelius, *Ann. Philosophy* **3**, 362 (1814).
3. J. B. Stas, *Nouvelles Recherches sur les Lois des Proportions Chimique* (1865).
4. J. C. G. de Marignac, *Compt. Rend.* **14**, 570 (1842).
5. T. W. Richards and R. C. Wells, *J. Am. Chem. Soc.* **27**, 459 (1905).
6. J. L. Meyer, *Liebigs Ann. Chem., Suppl. Band* **7**, 354 (1870).
7. D. I. Mendelyeev, *J. Russian Chem. Soc.* **1**, 60 (1869); *Z. Chem.* **12**, 405 (1869).
8. B. Taylor, *Phil. Trans.* xxxii, 291 (1723).
9. P. L. Dulong and A. T. Petit, *Ann. Chim.* **10**, 395 (1819).
10. I. Newton, *Principia* Bk. ii, prop. 23 (1687).
11. J. Dalton, *A New System of Chemical Philosophy* (1808).
12. L. J. Gay-Lussac, *Mém. Soc. Arcueil* **2**, 207 (1809).
13. A. Avogadro, *J. Phys.* lxxiii, 58 (1811).
14. A. M. Ampere, *Ann. Chim. et Phys.* xc, 45 (1814).
15. P. L. Dulong and A. T. Petit, *Ann. Chim. et Phys.* **10**, 395 (1819).
16. E. Mitscherlich, *Abhandl. Kgl. Akad. Wiss. Berlin* 427 (1820).
17. J. B. Dumas, *Ann. Chim. et Phys.* **33**, 337 (1826).
18. M. Faraday, *Phil. Trans. Roy. Soc. London* 77 (1834).
19. S. Cannizzaro, *Nuovo Cimento* **7**, 321 (1858).
20. L. J. Gay-Lussac, *Ann. Chim.* xliii, 137 (1802).
21. W. Thomson, *Cambridge Phil. Soc. Proc.* I, 66 (1866) [1848].
22. L. J. Gay-Lussac, *Ann. Chim. et Phys.* **21**, 330 (1822).
23. H. V. Regnault, *Compt. Rend.* **69**, 780 (1869).
24. A. Volta, *Phil. Trans. Roy. Soc. London* 403 (1800).
25. M. Faraday, *Experimental Researches in Electricity* Vol. I, Bernard Quaritch (London), 1839.
26. S. E. Virgo, *Science Progress* **27**, 634 (1933).
27. R. T. Birge, *Am. J. Phys.* **13**, 63 (1945).
28. L. C. King and E. K. Neilsen, *J. Chem. Ed.* **35**, 198 (1958).
29. H. Cavendish, *Phil. Trans. Roy. Soc. London* **56**, 141 (1766).
30. Richards and Baxter, *Proc. Am. Acad.* **35**, 253 (1900).
31. A. Werner, *Z. Anorg. Chem.* **3**, 267 (1893); A. Werner and A. Miolati, *Z. Physik. Chem.* **12**, 35 (1893).
32. R. D. Foust and P. C. Ford, *J. Chem. Ed.* **47**, 165 (1970).
33. W. Prout, *Ann. Chim. et Phys.* [2], **10**, 373 (1818).

34. F. Wöhler, *Ann. Physik Chem.* **12**, 253 (1828).
35. J. von Liebig, *Liebigs Ann. Chem.* **38**, 110 (1841).
36. T. Clark, *Ann. Chim.* **XLI**, 276 (1829).
37. Fr. Rüdorff, *Poggendorfs Ann. Phys.* **114**, 63 (1861).
38. F. M. Raoult, *Compt. Rend.* **95**, 1030, 187; **94**, 1517 (1882).
39. L. Wilhelmy, *Poggendorfs Ann. Phys.* **81**, 413, 449 (1850).
40. M. Berthelot and P. de St. Gilles, *Ann. Chim. et Phys.* **65**, 385 (1862).
41. C. M. Guldberg and P. Waage, *J. Prakt. Chem.* [2], **19**, 69 (1879); also *Etude sur les Affinites Chimiques* Impr. de Brogger & Christie, Christiania, 1867.
42. Jones and Lapworth, *J. Chem. Soc.* 1427 (1911).
43. R. Clausius, *Poggendorfs Ann. Phys.* **101**, 338 (1857).
44. F. Kohlrausch, *Göttinger Nachrichten* 213 (1876); *Poggendorfs Ann. Phys.* **6**, 1, 145 (1879).
45. J. H. van't Hoff, *Z. Physik. Chem.* **1**, 481 (1887), and earlier papers.
46. S. Arrhenius, *Z. Physik. Chem.* **1**, 631 (1887).
47. C. B. Anderson and Stanley E. Wood, *J. Chem. Ed.* **42**, 658 (1965).
48. Lambert, *Photometria Sive de Mensura et Gradibus Luminis, Colorum et Umbrae* (1760).
49. Beer, *Ann. Phys.* **86**, 78 (1852).
50. P. Bouguer, *Essai d'Optique sur la Gradation de la Lumière* (1729).
51. Bunsen and Roscoe, *Ann. Phys.* **101**, 235 (1857).
52. J. Volhard, *J. Prakt. Chem.* **117**, 217 (1874).
53. Nernst, *Z. Physik. Chem.* **4**, 129 (1889).
54. J. C. Poggendorf, *Poggendorfs Ann. Phys.* **56**, 324 (1842).
55. Stanley E. Wood and C. B. Anderson, *J. Chem. Ed.* **42**, 659 (1965).
56. J. F. Daniell, *Phil. Trans. Roy. Soc. London* **126**, 107 (1836).
57. J. H. van't Hoff, *Étude de Dynamique Chimique* F. Muller & Co., Amsterdam (1884).
58. J. J. Hood, *Phil. Mag.* **20**, 323 (1885).
59. S. Arrhenius, *Z. Physik. Chem.* **4**, 226 (1889).

APPENDIX 2/PREPARATION OF SOLUTIONS

Acetic acid ($C_2H_4O_2$, 60 g mole^{-1}), 6.0M: Add water to 343 ml of glacial acetic acid to make 1.0 liter of solution.

Aluminum chloride ($AlCl_3 \cdot 6H_2O$, 241.4 g mole^{-1}), 0.05M: Dissolve 12.1 g of $AlCl_3 \cdot 6H_2O$ in water to make 1 liter of solution.

Ammonia (called ammonium hydroxide) (NH_3, 17.0 g mole^{-1}), 6.0M: Dissolve 810 ml of reagent concd. ammonia solution (28-30% assay, sp gr 0.90) in water, and dilute to make 1.0 liter of solution. 1.0M: Dissolve 135 ml of concd. ammonia solution (28-30% assay, sp gr 0.90) in water, and dilute to make 1.0 liter of solution.

Ammonium carbonate, 1.0M: Dissolve 96.1 g of ammonium carbonate [$(NH_4)_2CO_3$, 96.1 g mole^{-1}] in water, and dilute to make 1 liter of solution.

Ammonium chloride, 1.0M: Dissolve 53.5 g of ammonium chloride (NH_4Cl, 53.5 g mole^{-1}) in water, and dilute to make 1 liter of solution. 0.01M: Dissolve 0.54 g of ammonium chloride in 1 liter of water, or dilute 1.0M solution 1:100.

Ammonium nitrate, 1.0M: Dissolve 80.1 g of ammonium nitrate (NH_4NO_3, 80.1 g mole^{-1}) in water, and dilute to make 1.0 liter of solution.

Arsenic trichloride, 0.02M: Dissolve 3.6 g of arsenic trichloride ($AsCl_3$, 181.3 g mole^{-1}) in water, and dilute to make 1 liter of solution. If a cloudy solution develops, add hydrochloric acid to make it clear.

Barium nitrate, 0.1M: Dissolve 26.1 g of barium nitrate [$Ba(NO_3)_2$, 261.4 g mole^{-1}] in water, and dilute to make 1 liter of solution.

Bromthymol blue indicator: Dissolve 0.4 g of dibromothymolsulfonphthalein in 6 ml of 0.1M sodium hydroxide, and dilute to 1 liter of solution with water.

Buffer solutions

pH 1: 0.10M hydrochloric acid.

pH 2: 0.010M hydrochloric acid.

pH 3: 1000 ml of 0.1M potassium hydrogen phthalate and 446 ml of 0.1M hydrochloric acid.

pH 4: 1000 ml of 0.1M potassium hydrogen phthalate and 20 ml of 0.1M hydrochloric acid.

pH 5: 1000 ml of 0.1M potassium hydrogen phthalate and 452 ml of 0.1M sodium hydroxide.

pH 6: 1000 ml of 0.1M potassium dihydrogen phosphate and 112 ml of 0.1M sodium hydroxide.

pH 6.5: 1000 ml of 0.1M potassium dihydrogen phosphate and 278 ml of 0.1M sodium hydroxide.

pH 7: 1000 ml of 0.1M potassium dihydrogen phosphate and 582 ml of 0.1M sodium hydroxide.

pH 7.5: 1000 ml of 0.1M potassium dihydrogen phosphate and 818 ml of 0.1M sodium hydroxide.

pH 8: 1000 ml of 0.025M sodium tetraborate (borax) and 410 ml of 0.1M hydrochloric acid.

pH 9: 1000 ml of 0.025M sodium tetraborate and 92 ml of 0.1M hydrochloric acid.

pH 10: 1000 ml of 0.025M sodium tetraborate and 366 ml of 0.1M sodium hydroxide.

pH 11: 1000 ml of 0.05M sodium bicarbonate and 454 ml of 0.1M sodium hydroxide.

pH 12: 1000 ml of 0.05M sodium monohydrogen phosphate and 538 ml of 0.1M sodium hydroxide.

pH 13: 0.1M sodium hydroxide.

pH 14: 1M sodium hydroxide.

Calcium nitrate, 0.02M: Dissolve 4.72 g of calcium nitrate tetrahydrate [$Ca(NO_3)_2 \cdot 4H_2O$, 236.2 g mole^{-1}] in water to make 1 liter of solution.

Cupric chloride, 0.10M: Dissolve 17.1 g of $CuCl_2 \cdot 2H_2O$ (170.5 g mole^{-1}) in water to make 1.0 liter of solution.

Cupric nitrate, 0.02M: Dissolve 4.83 g of $Cu(NO_3)_2 \cdot 3H_2O$ (241.6 g mole^{-1}) in water to make 1 liter of solution.

Cupric sulfate, 1.0M: Dissolve 249.7 g of $CuSO_4 \cdot 5H_2O$ (249.7 g mole^{-1}) in water, and dilute to make 1.0 liter of solution. 0.10M: Dissolve 24.97 g of $CuSO_4 \cdot 5H_2O$ in water to make 1.0 liter of solution.

Dimethylglyoxime ($C_4H_6N_2O_2$), 0.1%: Dissolve 1.0 g in 1 liter of ethyl alcohol.

Ethylenediamine, 10%: Mix 50 ml of ethylenediamine with 450 ml of distilled water. It is toxic. Wash well with water if spilled on clothes or skin. Do not make up more than a week ahead.

Ferric ammonium sulfate, saturated: Dissolve about 1240 g of ferric ammonium sulfate dodecahydrate ($FeNH_4SO_4 \cdot 12H_2O$, 482.2 g mole^{-1}) in 1 liter of water.

Ferric chloride, 0.02M: Dissolve 0.54 g of $FeCl_3 \cdot 6H_2O$ (270.3 g mole^{-1}) in water to make 1 liter of solution. 0.005M: Dissolve 0.14 g of hydrated salt to make 1 liter of solution. This last solution should be made freshly.

Hydrochloric acid, 6.0M: Add 490 ml of hydrochloric acid (assay 37-38%, sp gr 1.19) to about 500 ml of water (*caution*: heat given off) and dilute to make 1.0 liter of solution. 1.0M: Dissolve 81.67 ml of hydrochloric acid (assay 37-38%) in 500 ml of water and dilute to make 1.0 liter of solution.

Hydrogen peroxide, 0.2M: Add 1 liter of 30% hydrogen peroxide (*caution*: read label) to 43 liters of water. Store only in bottles with blow-out caps.

Lead nitrate, 0.05M: Dissolve 16.6 g of lead nitrate [$Pb(NO_3)_2$, 331.2 g $mole^{-1}$] in water, and dilute to make 1 liter of solution.

Mercuric nitrate, 0.02M: Dissolve 6.9 g of mercuric nitrate hydrate [$Hg(NO_3)_2 \cdot H_2O$, 342.6 g $mole^{-1}$] in water, and dilute to make 1 liter of solution.

Methyl orange indicator, 0.1%: Dissolve 1 g of the dye in 1 liter of water.

Methyl red indicator, 0.1%: Dissolve 1 g of dye in 37 ml of 0.1M sodium hydroxide, and dilute with water to make 1 liter of solution.

Methyl violet indicator, 0.25%: Dissolve 2.5 g of the dye in water to make 1 liter of solution.

Nickel nitrate, 0.02M: Dissolve 5.8 g of nickel nitrate hexahydrate [$Ni(NO_3)_2 \cdot 6H_2O$, 290.8 g $mole^{-1}$] in water, and dilute to make 1 liter of solution.

Nitric acid, 6.0M: Add 378 ml of concentrated reagent (assay 70.0-71.0%, sp gr 1.42) to 500 ml of water (*caution*: heat is evolved), and dilute to 1 liter of solution.

Phenolphthalein indicator, 0.1%: Dissolve 1 g of dye in 700 ml of ethyl alcohol, and dilute with water to make 1 liter of solution.

Potassium hydrogen phthalate, 0.10M: Dissolve 20.4 g of compound ($KHO_4C_8H_4$, 204.22 g $mole^{-1}$) in water, and dilute with water to make 1.0 liter of solution.

Potassium chromate, 0.5M: Dissolve 97.1 g of compound (K_2CrO_4, 194.2 g $mole^{-1}$) in water, and dilute to make 1 liter of solution.

Potassium dihydrogen phosphate, 0.10M: Dissolve 13.6 g of compound (KH_2PO_4, 136.1 g $mole^{-1}$) in water, and dilute to make 1.0 liter of solution.

Potassium ferrocyanide, 0.1M: Dissolve 42.2 g of potassium ferrocyanide trihydrate [$K_4Fe(CN)_6 \cdot 3H_2O$, 422.39 g $mole^{-1}$) in water, and dilute to make 1 liter of solution.

Potassium hydroxide, 20%: Dissolve 250 g of potassium hydroxide pellets (85% assay) in 800 ml of water. *Caution*: heat is evolved. **1.0M**: Dissolve 65.9 g of pellets (85% assay) in water and dilute to make 1.0 liter of solution.

Potassium thiocyanate, 0.1M: Dissolve 9.7 g of potassium thiocyanate (KSCN, 97.18 g $mole^{-1}$) in water, and dilute to 1 liter of solution. This solution will not keep indefinitely, owing to bacterial action.

Silver nitrate, 1.0M: Dissolve 169.9 g of silver nitrate ($AgNO_3$, 169.9 g $mole^{-1}$) in water, and dilute to make 1.0 liter of solution. **0.02M**: Dissolve 3.4 g of silver nitrate in water, and dilute to make 1 liter of solution.

Sodium acetate, 1.0M: Dissolve 82.0 g of sodium acetate ($NaC_2H_3O_2$, 82.04 g mole^{-1}) in water, and dilute to make 1.0 liter of solution.

Sodium bicarbonate, 0.05M: Dissolve 4.2 g of sodium bicarbonate ($NaHCO_3$, 84.02 g mole^{-1}) in water, and dilute to make 1 liter of solution.

Sodium chloride, 1.0M: Dissolve 58.5 g of sodium chloride (NaCl, 58.45 g mole^{-1}) in water, and dilute to make 1.0 liter of solution. A 5% solution is almost the same concentration as 1.0M, but may be made also by dissolving 50 g of NaCl in 950 ml of water.

Sodium hydrogen phosphate, 0.050M: Dissolve 13.4 g of sodium hydrogen phosphate ($Na_2HPO_4 \cdot 7H_2O$, 268.1 g mole^{-1}) in water, and dilute to make 1.0 liter of solution.

Sodium hydroxide, 6.0M: Dissolve 247 g of sodium hydroxide pellets (assay 97%, NaOH, 40.0 g mole^{-1}) in water (*caution*: heat is evolved), and dilute to make 1.0 liter of solution. 1.0M: Dissolve 41.2 g of the same NaOH pellets in water and dilute to make 1.0 liter of solution.

Sodium indigodisulfonate, 0.1%: Dissolve 1 g of dye in 500 ml of ethyl alcohol and dilute with water to make 1 liter of solution.

Sodium nitroprusside, 10%: Dissolve 100 g of sodium nitroprusside ($Na_2[Fe(NO)(CN)_5] \cdot 2H_2O$, 297.97 g mole^{-1}) in 900 g of water.

Sodium tetraborate (borax), 0.025M: Dissolve 9.5 g of sodium tetraborate decahydrate ($Na_2B_4O_7 \cdot 10H_2O$, 381.4 g mole^{-1}) in water, and dilute to make 1.0 liter of solution.

Sulfuric acid, 3.0M (6.0N): Add 166 ml of concentrated reagent (assay 95.5-96.5%, sp gr 1.84) to 700 ml of water *cautiously*, and then dilute to make 1 liter of solution. A large amount of heat is given off, so it may be useful to use 700 ml of ice water. The molecular weight of H_2SO_4 is 98.08 g mole^{-1}.

Thymol blue, 0.04%: Dissolve 0.4 g of thymol blue in 8.6 ml of 0.1M sodium hydroxide and 35 ml of water. Dilute this to 1 liter with more water.

Thymolphthalein, 0.1%: Dissolve 1 g in 1 liter of ethyl alcohol.

APPENDIX 3/EQUIPMENT REQUIREMENTS

The following equipment, costing about $22 (list prices), fills most of the needs in this laboratory manual for one student.

1 Asbestos square, 5 × 5 inches
1 Beaker, 50-ml
1 Beaker, 100-ml
1 Beaker, 150-ml
1 Beaker, 250-ml
1 Beaker, 400-ml
1 Beaker, 600-ml
1 Clamp, pinch
1 Cylinder, graduated, 10-ml
1 Cylinder, graduated, 100-ml
1 Dish, evaporating, size 00A
2 Flasks, Erlenmeyer, 125-ml
2 Flasks, Erlenmeyer, 250-ml
1 Flask, filter, 500-ml
1 Funnel, Hirsch, size 4/0 or 3/0
1 Funnel, polyethylene, 75-mm
1 Holder, test tube
1 Medicine dropper
1 Pipette, Mohr graduated, 5-ml
1 Pipette, transfer, 5-ml
1 Pipette bulb, 30-ml
1 Rack, test tube
1 Spatula
1 Stirring rod
6 Test tubes, 13- × 100-mm
6 Test tubes, 18- × 150-mm
1 Test tube, 25- × 200-mm
1 Thermometer, −10°C to 110°C
1 Wash bottle, polyethylene squeeze, 250-ml
1 Watch glass, 65-mm
1 Watch glass, 100-mm
1 Wire gauze

General equipment available in the laboratory or storeroom includes:

Analytical balances of 1-mg readability or better

Triple beam balances

Barometer

Clock with second hand

Drying oven

House vacuum line or aspirators

Ring stands, clamps

Burners

Centrifuges

Burette

Filter crucible, sintered glass (Experiment 5)

Less frequently used equipment includes:

Gas thermometer (Experiment 8)

Colorimeter (Experiment 15, Part 6, and Experiment 24, Part 2). One is sufficient for 8 students.

Potentiometer (Experiment 25)

Conductance bridge (Experiment 22)

$0.1°C$ Thermometer, $-1°C$ to $50°C$ (Experiments 6, 19, 20)

Storage batteries, 6 V (Experiment 9)

Ammeter (Experiment 9)

APPENDIX 4/SAMPLE REPORT FOR ASSIGNMENT 8

p.3

4 October 1971

Data for Assignment 8
Barometer was 758 mm

Temp.	Manometer reading		Gas pressure
25°	+34 mm	(atmosphere side higher)	792
0°	−35	(atmosphere side lower)	723
50°	+100		858
75°	+159		917
100°	+235		993
Dry Ice	−217		541

p.4

4 October 1971

Report of Assignment 8: The Constant Volume Gas Thermometer

The relationship between pressure and temperature for a gas at constant volume was experimentally examined. From this relationship, the sublimation temperature of solid carbon dioxide was determined. A relationship between pressure and an absolute temperature scale was shown.

The apparatus used was essentially a glass bulb connected to a mercury manometer. The glass bulb containing the gas was immersed in a bath which was adjusted to various temperatures. The pressure difference between the atmosphere and the bulb was determined by measuring the difference in height of the two arms of the manometer. The barometric pressure was 758 mm of mercury. The data obtained are listed in Table 1.

Table 1

Temperature (°C)	Manometer reading (mm of Hg)	Gas pressure (mm of Hg)
100°	235	993
75°	159	917
50°	100	858
25°	34	792
0°	-35	723
Dry Ice	-217	541

The data of Table 1 are plotted in the graph on pages 6 and 7, and a straight line is drawn through the points. Extrapolation of the line indicates that the temperature at zero pressure is -272°C, the absolute

p. 5

zero. The pressure found for the gas at the sublimation temperature of Dry Ice corresponds to a temperature of $-68°C$ on the graph.

The error in the determination of absolute zero, which should be $-273°C$, is probably fortuitously small. The volume of gas outside the bulb should cause an error in pressure when the bulb temperature is very different from room temperature. The part of the confined gas outside the bulb was at room temperature. The extrapolation over 270° also should cause a large error. The error in the sublimation temperature of carbon dioxide, which should be $-80°C$, is substantial; it is probably due to incomplete cooling of the bulb, and perhaps the causes mentioned, which may not cancel at this temperature.

The temperatures in Table 1 are converted to an absolute temperature scale using the determined absolute zero $(-272°C)$ in Table 2.

Table 2

°C	°Absolute	Pressure (torr)	P/T
0	272	723	2.68
25	297	792	2.67
50	322	858	2.67
75	347	917	2.65
100	372	993	2.67

The ratio of pressure and this absolute temperature, P/T, is found to vary from 2.65 to 2.68, which is very nearly constant. The relationship between pressure and absolute temperature can be expressed by the equation $P/T = $ a constant.

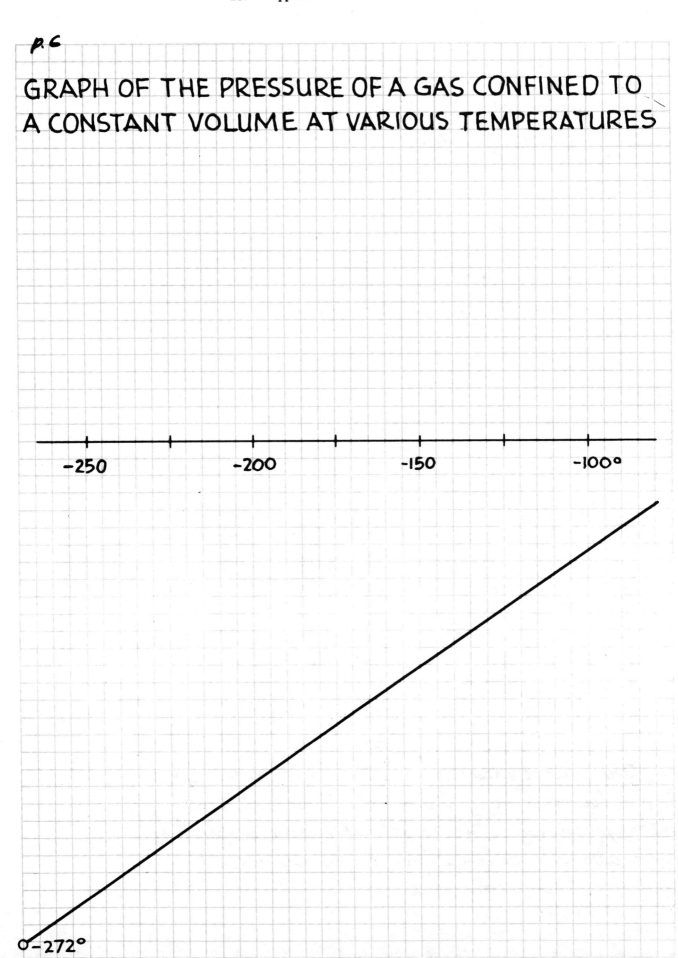

Appendix 4 233

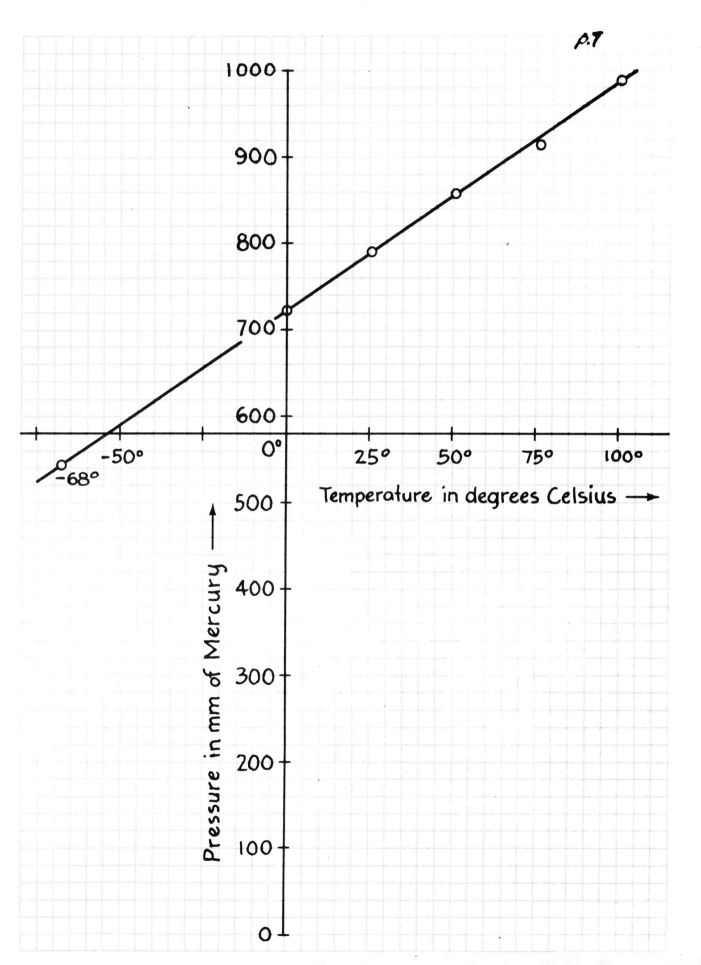